PLATED STRUCTURES

Stability and Strength

Related titles

AXIALLY COMPRESSED STRUCTURES: STABILITY AND STRENGTH

edited by R. Narayanan

BEAMS AND BEAM COLUMNS: STABILITY AND STRENGTH

edited by R. Narayanan

PLATED STRUCTURES

Stability and Strength

Edited by

R. NARAYANAN

M.Sc.(Eng.), Ph.D., D.I.C., F.I.Struct.E., F.I.C.E., F.I.E.
*Senior Lecturer, Department of Civil and Structural Engineering,
University College, Cardiff, United Kingdom*

CRC Press
Taylor & Francis Group
Boca Raton   London   New York

CRC Press is an imprint of the
Taylor & Francis Group, an **informa** business
A TAYLOR & FRANCIS BOOK

CRC Press
Taylor & Francis Group
6000 Broken Sound Parkway NW, Suite 300
Boca Raton, FL 33487-2742

First issued in paperback 2019

© 2005 by Taylor & Francis Group, LLC
CRC Press is an imprint of Taylor & Francis Group, an Informa business

No claim to original U.S. Government works

ISBN-13: 978-0-85334-218-2 (hbk)
ISBN-13: 978-0-367-87161-1 (pbk)

British Library Cataloguing in Publication Data

Plated structures.
  1. Plates, Iron and steel   2. Structural
  engineering
  I. Narayanan, R.
  624.1'821   TA684

WITH 10 TABLES AND 138 ILLUSTRATIONS

Visit the Taylor & Francis Web site at
http://www.taylorandfrancis.com

and the CRC Press Web site at
http://www.crcpress.com

PREFACE

I have great pleasure in writing a short preface to this book on *Plated Structures*, the third of the planned set of volumes on the stability and strength of structures.

The collapse a few years ago of four large box girders during erection precipitated a considerable research effort in many countries on the various design aspects of plated structures (such as box girders and plate girders). Revisions of design codes currently taking place in many countries are based on the results of fundamental research, aimed at an improved understanding of the complex problems of stability. The object of this book is to explain the current theories, and to provide the theoretical background to the specifications. Many of these theories have, in fact, been incorporated in recent design codes.

As the book is addressed to structural designers and post-graduate students, a fundamental knowledge of structural mechanics is taken for granted; nevertheless, sufficient introductory material is included in each chapter to make the subject matter easily readable.

This volume contains eight chapters, all of which are written by persons who have made notable contributions to the relevant subject area. The first four chapters are devoted to various aspects of the design of a plate girder and concentrate largely on the associated stability problems in shear. The four remaining chapters are largely concerned with stability problems in compression, such as those met in box girder flanges and ship hulls. Chapter 8 also highlights some of the complexities of interaction between different stresses. Thus the book covers a wide range of topics of relevance to both the designer and the student.

I am very grateful to all the contributors for the willing cooperation they extended in producing this volume. I sincerely hope that the book will meet the needs of the researcher and the engineer alike.

R. NARAYANAN

CONTENTS

LIST OF CONTRIBUTORS

H. R. EVANS

Reader, Department of Civil and Structural Engineering, University College, Newport Road, Cardiff CF2 1TA, UK.

J. E. HARDING

Lecturer, Department of Civil Engineering, Imperial College of Science and Technology, Imperial College Road, London SW7 2BU, UK.

V. KŘÍSTEK

Professor, Research Consultant, Faculty of Civil Engineering, Czech Technical University, Prague, Thakur Street 7, 166 29 Prague 6, Czechoslovakia.

N. W. MURRAY

Professor of Structural Engineering, Department of Civil Engineering, Monash University, Clayton 3168, Victoria, Australia.

R. NARAYANAN

Senior Lecturer, Department of Civil and Structural Engineering, University College, Newport Road, Cardiff CF2 1TA, UK.

T. M. ROBERTS

Lecturer, Department of Civil and Structural Engineering, University College, Newport Road, Cardiff CF2 1TA, UK.

N. E. SHANMUGAM

Senior Lecturer, Department of Civil Engineering, National University of Singapore, Kent Ridge, Singapore 0511.

M. ŠKALOUD

Professor, Institute of Theoretical and Applied Mechanics, Czechoslovak Academy of Sciences, Vysehradska 49, 128 49 Prague 2, Czechoslovakia.

Chapter 1

LONGITUDINALLY AND TRANSVERSELY REINFORCED PLATE GIRDERS

H. R. EVANS

Department of Civil and Structural Engineering,
University College, Cardiff, UK

SUMMARY

To maximise the strength/weight ratio of a plate girder, advantage must be taken of the post-buckling capacity: this capacity is considered in the present chapter. Derivation of fundamental theory is included, together with a discussion of how such theory can be applied in design. Both transversely and longitudinally stiffened webs are considered with attention being paid to the design of the stiffeners themselves as well as to the overall collapse behaviour of the girder.

NOTATION

b Clear width of web plate between stiffeners
b_f Width of flange plate
c Position of plastic hinge
d Clear depth of web plate between flanges
E Elastic modulus
M_F Plastic moment of resistance provided by flanges
M_{pf} Plastic moment of the flange plate
M_p^* Flange strength parameter $= M_{pf}/d^2 t\sigma_{yw}$
M_u Maximum moment that can be carried by the girder
t Thickness of web plate
t_f Thickness of flange plate

1

V_s Ultimate shear capacity of girder
V_{yw} Shear force to produce yielding of web $= \tau_{yw}dt$

θ Inclination of web tension field
v Poisson's ratio
σ_t^y Tension field web membrane stress
σ_{yw} Yield stress of web material
σ_{yf} Yield stress of flange material
τ_{cr} Critical shear stress of web
τ_{yw} Shear yield stress of web material
γ Parameter defining rigidity of web stiffener

All other symbols are defined as they first appear in the text.

1.1 INTRODUCTION

A fabricated plate girder, such as that shown diagrammatically in Fig. 1.1, is normally used to support vertical loads over long spans where the high bending moments and shearing forces developed exceed the capacity of the available hot-rolled universal beam sections. In the fabricated plate girder section, the primary function of the top and bottom flange plates is to resist the axial tensile and compression forces arising from the bending action, whilst the web plate resists the shear force—the longitudinal fillet welds connecting the component plates ensure the transference of longitudinal shear from the web to the flanges.

For a given applied bending moment, the axial forces in the flanges decrease as the web depth (d) is increased so that, from this point of view, it is advantageous to make the webs as deep as possible. To reduce the self-weight of the girder, the web thickness (t) is usually limited, with the consequence that the webs are normally of slender proportions (web proportions are normally expressed in terms of the slenderness ratio d/t).

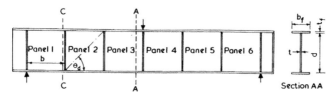

FIG. 1.1.   Elevation of a typical simply-supported plate girder.

The webs will then buckle at relatively low values of the applied shear loading.

In order to provide an efficient and economical girder design, advantage must be taken of the post-buckling capacity of the girder, i.e. its ability to carry loads considerably in excess of that at which the web buckles. In this way a girder of high strength/weight ratio can be designed, suitable for use in situations where reduction of self-weight is of prime importance. Examples of such situations commonly arise in long-span bridges, in aircraft and ship construction and in many other structures.

Whereas the determination of the initial buckling load of web plates has received much attention over many years, the study of the post-buckling behaviour is of more recent origin. This chapter will consider this post-buckling behaviour in some detail.

The case of transversely stiffened girders subjected to shear loading will be considered in the first instance, a collapse mechanism will be defined and an expression for the ultimate shear capacity will be derived. The influence of bending moments upon the shear capacity will then be considered and typical experimental results will be included to verify the proposed procedures.

The proposed approach assumes that the transverse stiffeners themselves remain effective right up to the failure load of the girder. The loads imposed upon these stiffeners in the post-buckling range will thus be studied in some detail and a procedure for the design of the stiffeners will be outlined.

Finally, girders with longitudinal, as well as transverse, web stiffeners will be considered and the basic analytical procedure will be slightly modified to take the longitudinal stiffeners into account. Experimental verification will again be included and some consideration will be given to the design of the stiffeners.

Because of the limitations on chapter length, all the equations and formulae presented cannot be derived explicitly. However, those equations of fundamental importance to the proposed method will be derived and, in all other cases, full reference to the appropriate source will be made. Throughout, the application of the analytical methods to the design of plate girders will be kept very much in mind, and in some instances, semi-empirical relationships will be presented for simplification.

1.2   TRANSVERSELY STIFFENED GIRDERS

The webs of plate girders are usually stiffened transversely, as shown in Fig. 1.1, to increase their shear carrying capacity. The spacing of these stiffeners (b) has a significant influence upon the girder behaviour and the aspect ratio (b/d) of the web panels is normally used as one parameter to express girder proportions.

Obviously, in a long-span girder such as that shown in Fig. 1.1, the various web panels will be subjected to different combinations of bending moments and shearing forces. For the particular case illustrated, panels close to the supports (1 and 6) will be subjected primarily to shear, whereas the central panels (3 and 4) will be subjected to high bending moments as well.

In this section of the chapter, attention will be concentrated, first of all, upon the shear capacity of a transversely stiffened girder. The basic theory for the prediction of the shear capacity will be discussed in some detail and the different phases of girder behaviour leading up to failure will be described. Then, the influence of bending moments upon the shear capacity will be considered and the necessary modifications to the theory will be outlined; this will be followed by a discussion of the bending capacity.

The parts of the theory relevant for a design analysis will then be summarised and the section will conclude with a comparison of some theoretical and experimental results for transversely stiffened girders.

1.2.1 Shear Capacity

(a) General

When a web plate is subjected to shear, then before it buckles, equal tensile and compressive principal stresses will be developed within the plate, as shown in Fig. 1.2(a). If the applied loading is increased further so that the plate buckles, it then loses its capacity to carry any additional compressive loading. In this post-buckled range, a new load carrying mechanism is developed within the plate, whereby any additional shear loading is carried by an inclined tensile membrane stress field, as shown in Fig. 1.2(b).

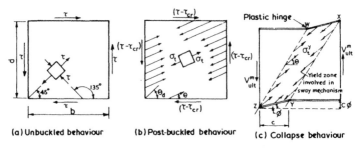

FIG. 1.2.   Phases in girder behaviour up to collapse.

This post-buckling behaviour was first noted by Wagner (1929) for girders of the proportions usually used in aircraft construction, i.e. girders with very slender webs and rigid flanges. For this type of girder, Wagner established expressions for the magnitude and inclination of the tensile membrane field. Further studies were carried out, for example by Kuhn and Peterson (1947) and Kuhn *et al.* (1962), to develop design methods for aircraft structures utilising the post-buckling reserve of strength.

These methods could not, however, be applied directly to the type of girders normally used in civil engineering because the girder proportions differ significantly. In particular, the flanges of civil engineering girders are usually much less rigid than those of aircraft girders, so that significant flange distortions can occur under the action of the forces imposed upon the flanges by the tensile membrane field (see Fig. 1.2(b)). These flange distortions influence the magnitude and inclination of the tension field developed in the web.

The first attempt to establish a method to predict the ultimate loads of girders of civil engineering proportions was made by Basler (1960). Basler assumed, rather conservatively, that the flanges were so flexible that they were not capable of withstanding the loads imposed upon them by the web tension field, so that the membrane stress field would then anchor against the vertical edges of the panel only. This assumption of very flexible flanges made by Basler represented the other end of the range from the very rigid flanges that had been assumed in the Wagner theory.

Rockey and Škaloud (1968) showed that for plate girders of the proportions normally employed in civil engineering, the flange rigidity was significant and had a great influence upon the ultimate load capacity. They established that the collapse mode of a plate girder involved the development of plastic hinges in the tension and compression flanges; collapse mechanisms of a similar nature were proposed by Fujii (1968), Chern and Ostapenko (1969), Komatsu (1971) and Calladine (1973).

All these proposed mechanisms differed from one another in some way and were limited in their range of application. A more generally applicable mechanism was proposed by Rockey *et al.* (1974) and after exhaustive comparisons with test data obtained from various sources the accuracy of this method has been established. Further discussion in this chapter will, therefore, be related specifically to this particular mechanism, which is illustrated in Fig. 1.2(c).

The method assumes that failure will occur when a certain region of the web yields as a result of the combined effect of the inclined tensile membrane stress field and the web buckling stress and when four plastic hinges form in the flanges. The resulting collapse mechanism then allows a shear sway displacement to develop, as illustrated diagrammatically in Fig. 1.2(c). Figure 1.3 shows a photograph of a large-scale girder recently tested by Evans and Tang (1981) and the similarity between the proposed and observed failure modes is apparent.

The behaviour of a plate girder under an increasing applied shear loading is represented by the three phases illustrated in Figs. 1.2(a), (b) and (c). Further consideration will now be given to each individual phase.

(b) Unbuckled Behaviour—Fig. 1.2(a)

Prior to buckling, if a uniform shear stress of magnitude τ is applied to the web, a principal tensile stress of magnitude τ will be set up at an inclination of 45° to the flange and a principal compressive stress of equal magnitude will be developed at an inclination of 135°. This stress system will exist until the applied shear stress (τ) reaches the critical value (τ_{cr}) at which the panel will buckle.

The value of the critical shear stress of an isolated panel can be determined from classical stability theory; the method has been clearly described in many texts, for example the recent book by Allen and Bulson (1980), and will not be repeated here. The critical shear stress depends upon the boundary conditions of the isolated panel but the true boundary conditions for a girder web are difficult to establish accurately because the degree of restraint imposed by the flanges and by the adjacent web panels cannot be evaluated. However, it can be assumed conservatively that all the

FIG. 1.3.   Collapse of large scale girder tested by Evans and Tang (1981).

boundaries of the web panel are simply supported, so that the critical shear stress is then given by

$$\tau_{cr} = k \, \frac{\pi^2 E}{12(1 - v^2)} \left(\frac{t}{d}\right)^2 \qquad (1.1)$$

where the buckling coefficient k is obtained as

$$k = 5 \cdot 35 + 4\left(\frac{d}{b}\right)^2 \qquad \text{when } \frac{b}{d} \geq 1$$

i.e. for wide panels; and

$$k = 5 \cdot 35 \left(\frac{d}{b}\right)^2 + 4 \qquad \text{when } \frac{b}{d} \leq 1$$

i.e. for webs with closely spaced transverse stiffeners.

(c) Post-buckled Behaviour—Fig. 1.2(b)
Once the web plate has lost its capacity to sustain any further increase in compressive stress, a new load carrying mechanism is developed. Additional loads, beyond the buckling load, are supported by a tensile membrane field, which anchors against the top and bottom flanges and against the adjacent members on either side of the web. The magnitude of this tensile membrane stress is indicated by σ_t in Fig. 1.2(b) and the angle of inclination of the membrane stress field, which is an unknown at this stage of the analysis, is denoted by θ.

Thus, the total state of stress in the web plate at this stage may be obtained by superimposing the post-buckled membrane stress upon those stresses set up when the applied shear stress reached its critical value τ_{cr}. By resolving these stresses in the direction along and perpendicular to the inclination θ, as shown in Fig. 1.4, the state of stress may be defined as

$$\sigma_\theta = \tau_{cr} \sin 2\theta + \sigma_t$$
$$\sigma_{(\theta + 90)} = -\tau_{cr} \sin 2\theta \qquad (1.2)$$
$$\tau_\theta = -\tau_{cr} \cos 2\theta$$

Since the flanges of the girder are of finite rigidity, they begin to bend inwards under the pull exerted by the tension field.

(d) Collapse Behaviour—Fig. 1.2(c)
Upon further increase of the applied loading, the tensile membrane stress (σ_t) developed in the web increases and a greater pull is exerted upon the

H. R. EVANS

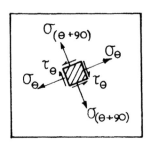

FIG. 1.4.   State of stress in web in post-buckling phase.

flanges. Eventually, the membrane stress reaches such a value that, when combined with the buckling stress as in eqn (1.2), the resulting stress (σ_θ) reaches the yield value (σ_{yw}) for the web material. This value of the membrane stress will be denoted as σ_t^y and it may be determined by applying the Von Mises–Hencky yield criterion:

$$\sigma_\theta^2 + \sigma_{(\theta+90)}^2 - \sigma_\theta\sigma_{(\theta+90)} + 3\tau_\theta^2 = \sigma_{yw}^2$$

By substituting the stresses from eqn (1.2), the value of the membrane stress to produce yield is obtained in terms of the critical buckling stress (τ_{cr}) and the inclination (θ) of the tension field:

$$\frac{\sigma_t^y}{\sigma_{yw}} = \sqrt{\left(1 - \left(\frac{\tau_{cr}}{\tau_{yw}}\right)^2 (1 - \tfrac{3}{4}\sin^2 2\theta)\right)} - \frac{\sqrt{3}}{2}\frac{\tau_{cr}}{\tau_{yw}}\sin 2\theta \qquad (1.3)$$

The equation is presented in non-dimensional form, where $\tau_{yw} = \sigma_{yw}/\sqrt{3}$.

Once the web has yielded, final failure of the girder will occur when plastic hinges have formed in the flanges, as shown in Fig. 1.2(c). It is a minimum requirement that the region WXYZ of the web must yield before a mechanism can develop, although the yield zone may well spread outside this region.

The failure load may be determined by applying a virtual sway displacement to the girder in its collapse state, as shown in Fig. 1.2(c).

It is convenient to consider the yielded region WXYZ of the web plate to be removed and to replace its action upon adjacent flange and web regions by the tensile membrane stresses, as shown in Fig. 1.5(a). Those stresses acting on the stationary section WZ obviously do no work during the virtual displacement. Also, in the particular case of a girder with identical top and bottom flanges, in the pure shear case the distances WX and ZY to the hinge positions will be identical so that the work done by the stresses acting on the top flange is balanced by that done by the stresses acting on

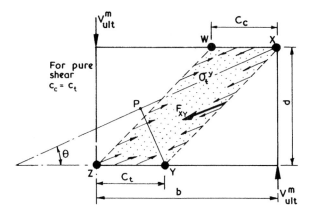

(a) Stresses exerted by yielded region

(b) Stresses & moments on comp. flange

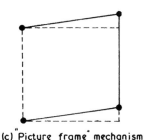

(c) "Picture frame" mechanism

FIG. 1.5.

the bottom flange. Thus, the only membrane stresses that will do any work during the virtual displacement are those acting on the face XY.

These stresses have a resultant force F_{xy}, as shown in Fig. 1.5(a), where

$$F_{xy} = \sigma_t^y t \sin \theta (d \cot \theta - b + c)$$

During the imposed virtual displacement shown in Fig. 1.2(c), the face XY will undergo an upward movement of magnitude $c\phi$. Thus, the external work done by the vertical component of the F_{xy} force (which acts in the downward direction) is given by

$$-F_{xy} \sin \theta c \phi$$

Further external work is done by the force V_{ult}^m, which is the post-buckling shear load that causes the mechanism to develop. As shown in Fig. 1.2(c), the work done in this case is equal to

$$V_{ult}^m c\phi$$

Thus, from these expressions:

$$\text{Total external work} = V_{ult}^m c\phi - F_{xy}\sin\theta c\phi$$
$$= V_{ult}^m c\phi - \sigma_t^y t\sin^2\theta(d\cot\theta - b + c)c\phi$$

From the Principle of Virtual Work, this external work must be balanced by the internal work done at the four plastic hinges. The internal work is obtained simply as

$$4M_{pf}\phi$$

where M_{pf} is the plastic moment capacity of the flange, which for a flange plate such as that shown in Fig. 1.1 is given simply by

$$M_{pf} = 0·25 b_f t_f^2 \sigma_{yf}$$

where σ_{yf} is the yield stress of the flange material.

Thus, by equating the internal and external work done, the post-buckling shear capacity V_{ult}^m is given by

$$V_{ult}^m = \sigma_t^y t\sin^2\theta(d\cot\theta - b + c) + \frac{4M_{pf}}{c}$$

To obtain the total failure load (V_s), the shear load ($\tau_{cr}dt$) required to make the web buckle in the first instance must be added on:

$$V_s = \tau_{cr}dt + \sigma_t^y t\sin^2\theta(d\cot\theta - b + c) + \frac{4M_{pf}}{c} \qquad (1.4)$$

This equation is not yet in a suitable form for calculating the ultimate load since it contains the term c, representing the position of the plastic hinges in the flanges, on the right-hand side; the hinge position is not yet known. It may be determined by considering the equilibrium of the flange and a free body diagram of the portion WX of the flange is shown in Fig. 1.5(b).

Since the internal plastic hinge will form at the point of maximum moment where the shear in the flange is zero, there will not be a lateral reaction at point W. Thus, taking moments about X:

$$2M_{pf} = \sigma_t^y t\frac{c^2}{2}\sin^2\theta$$

so that the hinge position is given by

$$c = \frac{2}{\sin\theta}\sqrt{\left(\frac{M_{pf}}{\sigma_t^y t}\right)} \tag{1.5}$$

Substituting this expression into eqn (1.4) and introducing a non-dimensional flange strength parameter M_p^* defined as

$$M_p^* = \frac{M_{pf}}{d^2 t \sigma_{yw}} \tag{1.6}$$

the ultimate load is obtained as

$$V_s = \tau_{cr}dt + \sigma_t^y t \sin^2\theta(d\cot\theta - b) + 4dt\sin\theta\sqrt{(\sigma_{yw}M_p^*\sigma_t^y)}$$

The equation may be non-dimensionalised by dividing throughout by the shear load required to produce yielding of the web ($V_{yw} = \tau_{yw}dt$). The final equation obtained is as follows:

$$\frac{V_s}{V_{yw}} = \frac{\tau_{cr}}{\tau_{yw}} + \sqrt{3}\sin^2\theta\left(\cot\theta - \frac{b}{d}\right)\frac{\sigma_t^y}{\sigma_{yw}} + 4\sqrt{3}\sin\theta\sqrt{\left(\frac{\sigma_t^y}{\sigma_{yw}}M_p^*\right)} \tag{1.7}$$

(e) Inclination of Tension Field
To calculate the ultimate shear load of any girder it is necessary to evaluate the critical shear stress (τ_{cr}) from eqn (1.1), the non-dimensional flange strength parameter (M_p^*) from eqn (1.6), and the tensile membrane stress (σ_t^y) required to produce yield from eqn (1.3) and then to substitute these values into eqn (1.7). However, there is one difficulty, since both eqn (1.7) and eqn (1.3) contain the term θ representing the inclination of the tension field and this is not yet known.

The value of θ cannot be determined directly and an iterative procedure has to be adopted in which successive values of θ are assumed and the corresponding ultimate shear load evaluated in each case. The process is repeated until the value of θ providing the maximum, and therefore the required, value of V_s has been established. The variation of V_s with θ is not very rapid and a parametric study carried out by Evans *et al.* (1976) established that for girders of normal proportions, the value of θ which produces the maximum value of V_s is approximately equal to 2/3 of the inclination of the diagonal of the web panel, i.e.

$$\theta \simeq \tfrac{2}{3}\tan^{-1}\left(\frac{d}{b}\right) \tag{1.8}$$

The assumption of this value of θ will lead either to the correct value or to an underestimation of the collapse load, so that it is a safe approximation. Equation (1.8) also gives a good starting value of θ if the process of iteration is to be carried out.

(f) Extreme Cases

From the expression derived in eqn (1.7), it is seen that the shear capacity is composed of three components. The first component, i.e. the first term on the right-hand side of the equation, represents the buckling strength; the second component indicates that part of the post-buckling tension field that is supported by the transverse stiffeners and the third component, which is a function of M_p^*, represents the contribution of the flanges.

In the case of a girder with very *weak flanges*, the value of the flange strength parameter M_p^* becomes very small; the third term then becomes negligible so that

$$\frac{V_s}{V_{yw}} = \frac{\tau_{cr}}{\tau_{yw}} + \sqrt{3} \sin^2 \theta \left(\cot \theta - \frac{b}{d} \right) \frac{\sigma_t^y}{\sigma_{yw}} \tag{1.9}$$

In such a case, none of the tension field is supported by the flanges and the field anchors completely against the transverse stiffeners. This is the situation assumed previously by Basler (1960), as discussed earlier, so that eqn (1.9) represents the true Basler solution.

When the *flanges are very strong*, then the distance of the plastic hinge away from the end of the panel (c) increases, as shown by eqn (1.5). When c becomes equal to the panel width (b), the hinges form at the four corners of the panel to form a 'picture frame' mechanism, as shown in Fig. 1.5(c), and the tension field develops at an inclination (θ) of 45°.

By substituting the value of $c = b$ into eqn (1.5) and taking $\theta = 45°$ in eqn (1.3), eqn (1.6) will give the limiting value of flange strength ($M_{p,lim}^*$) at which the switch to a picture frame mechanism will occur:

$$M_{p,lim}^* = \frac{1}{8} \left(\frac{b}{d} \right)^2 \left[\sqrt{\left(1 - \frac{1}{4} \left(\frac{\tau_{cr}}{\tau_{yw}} \right)^2 \right)} - \frac{\sqrt{3}}{2} \left(\frac{\tau_{cr}}{\tau_{yw}} \right) \right] \tag{1.10}$$

In such a case, substitution of the appropriate values of θ and σ_t^y/σ_{yw} into eqn (1.7), gives the following simplified expression for the ultimate shear capacity:

$$\frac{V_s}{V_{yw}} = \frac{1}{4} \frac{\tau_{cr}}{\tau_{yw}} + \frac{\sqrt{3}}{2} \sqrt{\left(1 - \frac{1}{4} \left(\frac{\tau_{cr}}{\tau_{yw}} \right)^2 \right)} + 4\sqrt{3} \left(\frac{d}{b} \right) M_p^* \tag{1.11}$$

Thus, when determining the ultimate shear load for any girder, the flange strength should be evaluated to see whether it exceeds the limiting value given in eqn (1.10). If so (and this will not often be the case for girders of civil engineering proportions), the expression for shear capacity given in eqn (1.11) should be used in place of the general expression given in eqn (1.7).

In the case of a girder with a *thick web*, then the web may yield before it buckles so that no tension field action will develop. Failure will then occur by a picture frame mechanism and substitution of $\tau_{cr} = \tau_{yw}$ into eqn (1.11) gives

$$\frac{V_s}{V_{yw}} = 1 + 4\sqrt{3}\left(\frac{d}{b}\right)M_p^*$$

By using the expression for M_p^* given in eqn (1.6), this may be further simplified to give the following equation for the shear capacity of a girder with an unbuckled web:

$$V_s = V_{yw} + \frac{4M_{pf}}{b} \tag{1.12}$$

A girder with a *very slender web* will, of course, have a low buckling capacity so that the ratio of τ_{cr}/τ_{yw} becomes small. As a consequence, there will be extensive post-buckling action and the value of the membrane stress (σ_t^y) will be high, as indicated by eqn (1.3). The general expression for the shear capacity given in eqn (1.7) will hold in such a case.

1.2.2 Webs Subjected to Coexistent Bending Moments and Shearing Forces
(a) General
When a girder web is subjected to a bending moment in addition to shear, the determination of the ultimate load capacity becomes more complex, as discussed by Evans *et al.* (1978). The interaction between the effects of shear and bending can be represented conveniently by the type of diagram shown in Fig. 1.6 where the shear capacity of the girder is plotted on the vertical axis and the bending moment capacity is plotted horizontally. Thus, any point on the interaction diagram shows the coexistent values of shearing force and bending moment that can be sustained by the girder.

The portion of the curve between points S and C represents the region within which the girder will fail by the development of a shear mechanism of the type discussed in Section 1.2.1. The vertical ordinate of point S represents the pure shear capacity, as given by eqn (1.7). This shear

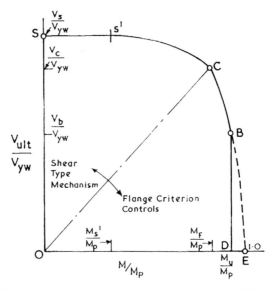

FIG. 1.6. Diagram showing interaction between shear and bending effects.

capacity will be reduced gradually by the presence of an increasing bending moment, as indicated in Fig. 1.6.

Beyond point C, where the applied bending moment is high, failure will occur in the flanges, either by yielding of the flange material, by inward buckling of the compression flange or by lateral buckling of the flange. The line OC on the interaction diagram, therefore, represents the dividing line between the shear mechanism type of failure and one of the flange failure modes.

A parametric study carried out by Evans *et al.* (1976) established that this change from a web to a flange failure mode normally occurs when the value of the applied bending moment is approximately equal to the plastic moment of resistance provided by the flange plates only, neglecting any contribution from the web. Thus, the horizontal coordinate at point C may be defined as being approximately equal to M_F where, for the typical cross-section illustrated in Fig. 1.1:

$$M_F = b_f t_f \sigma_{yf}(d + t_f) \tag{1.13}$$

This present section of the chapter will concentrate upon the region SC of the curve where failure occurs by the formation of a shear-type mechanism influenced by the presence of bending moments.

The presence of the bending moment requires three additional factors to be considered:

(a) the reduction in the buckling stress of the web due to the bending stresses;

(b) the influence of the bending stresses upon the magnitude of the membrane stresses required to produce yield in the web;

(c) the reduction in the plastic moment capacity of the flanges as a result of the axial flange stresses arising from the bending moment.

Each of these factors will now be discussed individually.

(b) Modified Web-buckling Stress
The modified shear-buckling stress (τ_{crm}) of the web due to the presence of bending stresses may be obtained from

$$(\tau_{crm}/\tau_{cr})^2 + (\sigma_{mb}/\sigma_{crb})^2 = 1 \tag{1.14}$$

where τ_{cr} is the critical stress for the pure shear case, as given by eqn (1.1), σ_{mb} is the compressive-bending stress in the extreme fibre at the mid-panel section, arising from the applied bending moment and σ_{crb} is the critical-buckling stress for the plate when it is subjected to a pure bending moment. This may again be determined from classical stability theory (see Allen and Bulson (1980) for example), and by assuming conservatively, as in the pure shear case, that the edges of the web are simply supported, the following expression is obtained:

$$\sigma_{crb} = 23 \cdot 9 \frac{\pi^2 E}{12(1-v^2)} \left(\frac{t}{d}\right)^2$$

(c) Modified Membrane Stresses for Web Yielding
In Sections 1.2.1(c) and 1.2.1(d), the value of the membrane stress (σ_t^y) required to be added to the critical buckling stress in order to produce yield in the web was established; the resulting expression was presented in eqn (1.3). The additional bending stress arising from the applied moment must also be taken into account now, and by applying the Von Mises–Hencky yield criterion, the following modified expression for the membrane stress is obtained:

$$\sigma_{tm}^y = -\tfrac{1}{2}A + \tfrac{1}{2}\sqrt{(A^2 - 4(\sigma_b^2 + 3\tau_{crm}^2 - \sigma_{yw}^2))} \tag{1.15}$$

where

$$A = 3\tau_{crm}\sin 2\theta + \sigma_b \sin^2\theta - 2\sigma_b \cos^2\theta$$

In this equation σ_b is the value of the bending stress which does, of course, vary over both the depth and width of the web panel; the corresponding value of σ_{tm}^y will, therefore, also vary over the web area. In the simpler pure shear case, the value of σ_t^y remained constant over the complete web so that the resultant (F_{xy}) of the membrane stresses, see Fig. 1.5(a), could be calculated directly. This is not possible when the membrane stress varies. In such a case, the resultant can only be determined by, first of all, evaluating σ_{tm}^y at a number of positions on the required section, taking the appropriate value of bending stress (σ_b) for each position; then from this known distribution of membrane stresses, the corresponding resultant can be calculated.

The magnitude of the membrane stresses will also vary along the junctions between the web and the flanges so that the equilibrium equations for the flanges (see Fig. 1.5(b) and eqn (1.5)) should be modified to take this into account. This may again be accomplished by evaluating σ_{tm}^y at a number of different positions at the flange/web junction and determining the resultant corresponding to the membrane stress distribution. However, experience has shown that for most girders, with little loss of accuracy but with a substantial simplification of the solution, the variation of the membrane stress along the flange/web junction can be neglected. A constant σ_{tm}^y value, calculated at the mid-point between hinges, may then be assumed for the portion of the flange lying between the hinges.

(d) Reduction in Plastic Moment Capacity of Flanges
In the shear case, the plastic moment capacity (M_{pf}) of the flanges considered in Section 1.2.1(d) was taken as the full value of $0.25 b_f t_f^2 \sigma_{yf}$. When high axial forces are developed in the flanges, as may well be the case when high bending moments are applied to the girder, then the effects of these axial forces in reducing the plastic moment capacity of the flange plates must be taken into account. From standard plasticity theory, the reduced capacity (M_{pf}') may be expressed as

$$M_{pf}' = M_{pf}\left[1 - \left(\frac{\sigma_f}{\sigma_{yf}}\right)^2\right] \tag{1.16}$$

where σ_f is the average axial stress for the portion of the flange between hinges.

This reduction equation must be applied to both the compression and tension flanges because the average axial stress developed within each flange will differ. It is also possible, of course, that the dimensions of the tension and compression flanges will differ so that the appropriate value of M_{pf} should be calculated for each flange.

(e) Solution Procedure

The shear sway failure mechanism is modified because of the effects of bending. Whereas, for the pure shear case, as shown in Fig. 1.5, the hinges in the flanges of a girder with identical top and bottom flanges formed at the same distance away from the end of the panel (i.e. the value of c for the two flanges was the same), this will not be so when bending effects are present, as shown in Fig. 1.7.

In fact, the additional considerations required to allow for the effects of bending make the determination of the ultimate load capacity considerably more complex. As in the pure shear case, the inclination (θ) of the membrane field is involved in the equations and this cannot be determined directly so that successive values of θ have to be assumed until the optimum value has been established. However, for each assumed value of θ, it is not now possible to determine the values of flange stresses directly so that a further iteration process is required.

In this secondary iterative procedure it is initially assumed that the average axial flange stress values (σ_f in eqn (1.16)) are zero; the corresponding flange hinge positions, defining the failure mechanism, and resulting ultimate load value are then calculated. More accurate flange stress values can then be calculated and the process is repeated. Successive

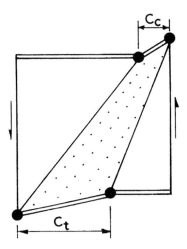

FIG. 1.7.   Shear failure mechanism influenced by bending ($C_c \neq C_t$).

cycles are carried out in which increasingly accurate values of flange stresses are assumed and the ultimate load value calculated in each case is compared to that obtained from the previous cycle. The iteration process is terminated when a sufficient degree of convergence has been achieved and the rate of convergence has been found to be very rapid in virtually all cases.

Once convergence has been achieved, this gives the ultimate load corresponding to the assumed value of membrane field inclination (θ). Successive values of θ then have to be taken to determine the optimum, as in the pure shear case.

The full solution procedure may, therefore, require several analyses to be carried out before the optimum values are established. Although, with experience, reasonably accurate starting values can be assumed to reduce the number of iterative cycles required, a 'hand' analysis is not really feasible and the procedure is most suitable for a computer solution. Indeed, the repetitive nature of the calculations enables a simple computer program to be written.

(f) Design Procedure

Rockey *et al.* (1978) presented a simplification of this procedure, suitable for use at the preliminary design stage. The simplified design procedure makes use of empirical relationships established from a detailed parametric study and enables the interaction diagram to be constructed directly, without recourse to a computer solution.

The parametric study established that the shear load capacity at point C on the interaction diagram, i.e. at the point where the failure mode switches from a shear mechanism to a flange failure, may be obtained approximately from the following empirical relationship:

$$\frac{V_c}{V_{yw}} = \frac{\tau_{cr}}{\tau_{yw}} + \frac{\sigma_t^y}{\sigma_{yw}} \sin 4 \frac{\theta d}{3} \left[0{\cdot}554 + 36{\cdot}8 \frac{M_{pf}}{M_F} \right] \left[2 - \left(\frac{b}{d}\right)^{1/8} \right] \quad (1.17)$$

This equation gives the vertical ordinate of point C and, taken in conjunction with the horizontal coordinate value, i.e. M_F as defined earlier in eqn (1.13), enables the position of point C to be plotted.

The parametric study also showed that the construction of the interaction diagram between points S and C could be simplified into two stages. Between points S and S' the curve remains horizontal, so that the vertical ordinate of point S' is the same as that of point S, i.e. the pure shear value given by eqn (1.7). Between points S' and C, the curve may be represented by a parabola, or for further simplification, with some underestimation of the failure load, by a straight line. The moment M'_S corresponding to the horizontal coordinate of point S' may be defined as $V_s b$, so that it becomes the maximum bending moment in the end panel of a simply supported girder, i.e. at a section such as CC in Fig. 1.1; the maximum allowable value of M'_S is restricted to $0.5 M_F$.

Thus by employing these empirical relationships, the portion SC of the interaction diagram, representing the region within which the girder fails by developing a shear mechanism, is defined directly.

1.2.3 Webs Subjected to Pure Bending

The region beyond point C of the interaction diagram plotted in Fig. 1.6, represents the range where the plate girder is subjected to high bending moments so that failure occurs in a 'bending', rather than in a 'shear' mode.

In a thin-walled girder, the portion of the slender web plate subjected to the compressive bending stresses will normally buckle, thereby losing its capacity to carry further compressive stresses. Thus, the distribution of longitudinal bending stresses over the depth of the web will become non-linear, as noted by Owen *et al.* (1970), and the web will shed some of the bending stresses that it should carry on to the flanges. The neutral axis of the cross-section will then move away from the compression flange, with a consequent increase in the longitudinal compressive strains developed within the flange.

The girder is thus unable to develop the full plastic moment of resistance (M_p) of the cross-section, calculated assuming a fully effective web. (It is this M_p value that is used as the denominator to non-dimensionalise the x-coordinate values of the interaction diagram.)

Provided that adequate lateral supports are provided to ensure that lateral buckling does not occur, then the girder will fail by inward collapse of the compression flange. This flange failure will occur at an applied moment value (M_u) which is approximately equal to the moment required to produce first yield in the extreme fibres of the compression flange. The required moment is, of course, reduced because of the effects of web buckling and it may be determined by applying an effective width formula, such as that proposed by Winter (1947), to the compression zone of the web and calculating the reduced section properties.

Although simple in principle, the resulting calculations are lengthy and time-consuming, but work reported by Cooper (1965 and 1971) enables the ultimate moment capacity to be determined directly by using an empirical formula. On the basis of an experimental study, Cooper proposed the following expression for the moment capacity:

$$\frac{M_u}{M_y} = 1 - 0 \cdot 0005 \frac{A_w}{A_f} \left[\frac{d}{t} - 5 \cdot 7 \sqrt{\frac{E}{\sigma_{yf}}} \right] \quad \text{for } M_u \not> M_p \quad (1.18)$$

where the term M_y in the denominator on the left-hand side of the equation represents the bending moment required to produce yield in the extreme fibre of the compression flange, assuming a fully effective web, i.e. neglecting the web-buckling effects.

The accuracy of this empirical formula has been confirmed by experiments carried out by various investigators. It enables the value of M_u, defining the position of point D on the horizontal axis of the interaction diagram to be determined directly.

This value of M_u is the moment required to produce yield in the extreme fibres of the flange and the corresponding stresses in the web will be below yield. Consequently, the web can support a certain amount of coexistent shear loading. This shear is defined by the ordinate of point B, lying vertically above point D, on the interaction diagram of Fig. 1.6. The ordinate V_B can be calculated simply, and with reasonable accuracy, as

$$\frac{V_B}{V_C} = \sqrt{\left(\frac{M_p - M_u}{M_{pw}} \right)} \quad (1.19)$$

where M_p is the full plastic moment capacity of the complete cross-section, as defined earlier, and M_{pw} is the plastic moment of resistance of the web plate by itself, i.e. $M_{pw} = 0 \cdot 25 t d^2 \sigma_{yw}$.

Equation (1.19) defines the position of the final point on the interaction diagram. The portion CB of the curve may be assumed to be parabolic, or for further simplification with little loss of accuracy, it may be assumed linear. In this way, the complete diagram can be drawn.

1.2.4 Experimental Verification for Transversely Stiffened Girders

The shear failure mechanism described in this chapter has been verified by extensive comparisons with test data obtained from various sources. These comparisons were summarised by Rockey et al. (1978); they considered the results of some 58 tests, carried out by various investigators, on transversely stiffened girders. The ratio between the predicted and experimentally measured values of failure loads for these 58 tests gave a mean value of 0·998, with a standard deviation of 0·06, thus proving the validity of the tension field approach.

More recently, Evans and Tang (1981) reported a series of five tests on a large-scale plate girder model. This model was specifically designed to be of extreme proportions so that it would provide a very stringent test of the theory. In particular, the web plate was very slender with a d/t ratio in excess of 800 to allow extensive post-buckling action to develop. Also, the transverse stiffeners were closely spaced to give a web-panel aspect ratio b/d in the region of 0·6.

The girder is shown diagrammatically in Fig. 1.8(a). The successive test panels were subjected to varying amounts of bending and shear and a comparison between the experimental points obtained and the predicted interaction diagrams is presented in Fig. 1.8(b). The agreement is observed to be excellent; indeed the maximum difference between the predicted and measured failure loads was only 2 %.

A view of the girder after completion of the test series is shown in Fig. 1.9. This gives an indication of the extensive buckling of the very slender web and of the large scale of the model girder. A close-up view of one of the test panels was shown earlier in Fig. 1.3 to illustrate the characteristics of a typical shear failure mechanism. An idea of the extent of the post-buckling action developed can be gained from the fact that the ultimate failure load in each test was observed to be about 30 times the load at which web buckling occurred.

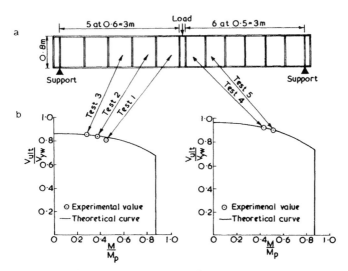

FIG. 1.8.   (a) Test girder; (b) comparison of experimental points and predicted interaction curves.

FIG. 1.9.   General view of girder after completion of five tests (Evans and Tang, 1981).

These recently obtained experimental results for a girder of extreme proportions give further confirmation of the validity of the shear failure mechanism approach and of its wide range of application.

1.3   DESIGN OF TRANSVERSE WEB STIFFENERS

1.3.1 General

Before post-buckling action can develop in a slender web, the boundary elements must be able to support the forces imposed upon them by the web tension field. The intermediate transverse web stiffeners, as shown in Fig. 1.1, are important load-carrying elements in this respect and the loading imposed upon them is complex. Until recently, little was known about the behaviour of these stiffeners when the web was operating in the post-buckled range.

Indeed, Bjorhovde (1980), in reporting a survey entitled 'Research Needs in Stability of Metal Structures' to the American Society of Civil Engineers, commented that: 'Present design methods for vertical stiffeners appear to be conservative. Research should be conducted to produce rational design rules for such elements ...'.

Rockey *et al.* (1981) established an ultimate load approach to the design of transverse stiffeners on webs loaded in shear. This approach was semi-empirical, being based partially upon experimental observations which enabled the loading imposed upon the stiffeners by the web tension field to be established. The method was considered further by Evans and Tang (1981); the stiffeners of the large test girder, shown earlier in Fig. 1.9, were instrumented so that the forces set up within them could be determined. This experimental study showed that the approach of Rockey *et al.* (1981) would lead to the design of safe, if slightly conservative, transverse stiffeners.

This section of the chapter will concentrate upon this particular approach to the design of transverse web stiffeners.

1.3.2 Behaviour of Transverse Stiffeners

The transverse web stiffeners play an important role in allowing the full ultimate load capacity of a girder to be achieved. In the first place, they increase the buckling resistance of the web; secondly, they must continue to remain effective after the web buckles to support the tension field; finally, they must prevent any tendency for the flanges to move towards one another.

The minimum rigidity that a transverse stiffener must possess to ensure that it remains straight and restricts buckling to the individual sub-panels of the web can be determined from linear buckling theory. This minimum rigidity is usually referred to as the γ^* value and tests have shown, for example those tests reported by Massonnet (1960), that it must be greatly increased to ensure that the stiffener remains effective when the web is loaded in the post-buckling range; indeed, values of up to $5\gamma^*$ have been recommended in some instances. Such empirical coefficients were included in several design codes, but with the move towards limit state design concepts in recent years the need for a more logical approach to stiffener design, based on the same theoretical model as that used to calculate the web shear capacity, became apparent.

1.3.3 Loads Imposed upon a Transverse Stiffener

Figure 1.10 illustrates a typical situation in which a transverse stiffener is positioned between two web panels, each of which has developed a shear failure mechanism; such a situation is the most critical from the point of view of designing the intermediate stiffener. The geometries of the two

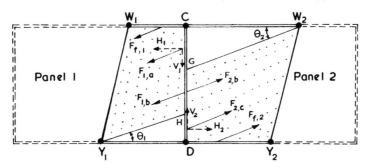

FIG. 1.10.   Forces imposed on transverse stiffener by tension fields (Rockey *et al.*, 1981).

mechanisms, i.e. the positions of the plastic hinges in the flanges and the inclination of the membrane stresses in the webs, can be determined as described in Sections 1.2.1(d) and 1.2.1(e) of this chapter. The tensile membrane stresses developed in the webs apply loads to the flanges and to the intermediate stiffener, as shown in the diagram.

The loads imposed upon the intermediate stiffener will now be determined. The resultant of the loads acting on the portion W_1C of the top flange is shown as $F_{f,1}$, acting at an inclination θ_1 to the horizontal. Similarly, the loads imposed upon the portion DY_2 of the lower flange have a resultant $F_{f,2}$, inclined at an angle θ_2. The vertical components of these two forces will tend to pull the flanges together and since the stiffener will resist this movement the components will be transmitted as end loads (V_C and V_D) to the stiffener. Thus:

$$V_C = -F_{f,1} \sin \theta_1 = -\sigma_{t,1}^y t W_1 C \sin^2 \theta_1 \qquad (1.20)$$

where the negative sign indicates that the force acts downwards and

$$V_D = F_{f,2} \sin \theta_2 = \sigma_{t,2}^y t DY_2 \sin^2 \theta_2 \qquad (1.21)$$

The loading imposed directly upon the stiffener by the web tension field can be separated into 3 zones. The top zone CG of the stiffener is subjected to the pull from the left-hand panel (panel 1) only, the resultant being designated as $F_{1,a}$. This resultant has a horizontal component H_1 which is assumed to be resisted by the wedge of web material CGW_2, and thus does not affect the stiffener, and a vertical component V_1 which forms an axial loading, where

$$V_1 = -\sigma_{t,1}^y t CG \sin \theta_1 \cos \theta_1 \qquad (1.22)$$

Similarly the bottom zone DH of the stiffener carries an axial loading V_2 as a result of the pull exerted by the tension field in the right-hand panel where

$$V_2 = \sigma_{t,2}^y t HD \sin \theta_2 \cos \theta_2 \qquad (1.23)$$

In the central zone GH of the stiffener, the force $F_{1,b}$ exerted by the left-hand panel is more or less balanced by the force $F_{2,b}$ from the right-hand panel so that the central region remains virtually unloaded by the tension field action.

The resulting forces exerted by the tension field upon the stiffener are then as shown in Fig. 1.11(a). In addition to the tension field loading, the stiffener is loaded over its whole length by the distributed forces arising from the differing shear stresses of the two adjacent panels; these forces are of intensity $(\tau_{cr,1} - \tau_{cr,2})t$ as shown in Fig. 1.11(b). The total axial forces imposed upon the stiffener are then as shown in Fig. 1.11(c).

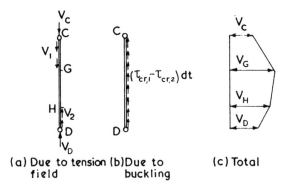

(a) Due to tension   (b)Due to   (c) Total
     field   buckling

FIG. 1.11.   Total axial forces imposed upon transverse stiffener.

Thus, once the ultimate shear loading and the geometry of the failure mechanism for the two adjacent web panels has been determined, eqns (1.20)–(1.23) give the magnitude and distribution of the axial loads that the transverse stiffener has to withstand. In Fig. 1.12, stiffener loading calculated in this way is compared to the loads actually measured by Evans and Tang (1981) in the tests on the large scale girder (see Fig. 1.9). The similarity between the measured and calculated load variation is apparent.

1.3.4 Analysis and Design of Stiffener
Having established the loads acting on the stiffener, the effects of these loads must be determined. The experimental study conducted by Rockey *et al.* (1981), and a similar study by Mele and Puthali (1979), has shown that a portion of the web plate acts with the stiffener in resisting the axial loading, despite the fact that the web is fully yielded by the tension field action. The effective cross-section of the stiffener is thus in the form of a T-shape, as shown in Fig. 1.13.

The exact amount of web acting with the stiffener is not known; indeed it is believed to vary as the shear mechanism develops. On the basis of their detailed experimental study, Rockey *et al.* (1981) proposed that a width of web of forty times its thickness (i.e. $40t$) should be empirically assumed to act with the stiffener.

The axial loading is assumed to be applied to the stiffener cross-section at the mid-thickness of the web plate; the true position of load application cannot be determined directly but the assumption made is conservative since it involves the greatest degree of eccentricity and thus causes the maximum possible bending effects for a given axial loading.

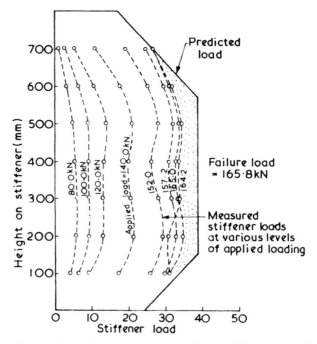

FIG. 1.12. Comparison of measured and predicted stiffener loads (Evans and Tang, 1981).

The stiffener is thus subjected to both axial loads (P) and bending moments. Before the bending moments can be evaluated, four factors must be considered:

(a) The moment arising due to the eccentricity of the applied load P from the centroidal axis of the stiffener. If this distance is defined as $\bar{x}$ then the corresponding moment is given by $P\bar{x}$.

(b) The moment due to any initial imperfection (δ_0) of the stiffener; this is given by $P\delta_0$.

(c) Since the stiffener is acting as a strut, the standard amplification factor $1/(1 - P/P_e)$ must be applied; P_e is the Euler buckling load for the stiffener strut, calculated considering the effective length to be the length GH, as defined in Fig. 1.10, i.e. the length over which the tension fields in the adjacent panels overlap.

(d) The final contribution to the bending effect is that associated with the disturbing action imposed by the buckled web upon the stiffener. The stiffness required to resist this is rather difficult to

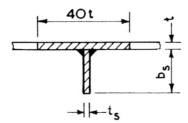

FIG. 1.13.   Effective section of transverse stiffener.

quantify but one possible method, which is followed in the recently published British Code of Practice (1982) for the design of steel bridges, is to apply a notional additional axial load to the stiffener to represent the de-stabilising action. This procedure was outlined by Chatterjee (1981) and the additional force may be taken as

$$\frac{4}{\pi^2}\frac{d^2}{b^2}\frac{t}{t_s}\tau_{cr} \times \text{effective area of stiffener cross-section}$$

The T-shaped stiffener cross-section is then designed as a strut to withstand the combined effects of axial loads and bending moments. A suitable expression was proposed by Horne (1979) to define the combination of axial load (P) and bending moment (M) that can be sustained by a member. For a typical stiffener section such as that shown in Fig. 1.13, the relationship takes the form:

$$\frac{M}{M_{ps}} = 1\cdot0 - \frac{\sigma_y t_s(b_s - \bar{x} + 0\cdot5t)^2}{M_{ps}}\frac{P}{P_s} \qquad (1.24)$$

where M_{ps} is the full plastic moment capacity of the section when there is no axial loading and P_s is the full axial yield, or squash, load of the section.

This relationship may be plotted in the form of an interaction diagram for the stiffener, as shown in Fig. 1.14, where the curve for a typical stiffener of the large girder tested by Evans and Tang (1981) has been plotted.

For any girder, the axial load to which the stiffener is subjected can be calculated according to the method described in Section 1.3.3. Then, provided that the coexistent moment is less than the allowable moment defined by eqn (1.24), i.e. provided the corresponding point plotted on the interaction diagram lies within the curve, the stiffener will be able to support the loads to which it is subjected.

Figure 1.14 shows the design value to lie just within the interaction diagram and this particular stiffener did, in fact, prove to be completely

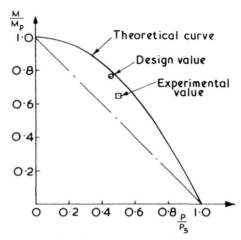

FIG. 1.14.   Interaction diagram for a typical transverse stiffener.

adequate in the test. The experimental point plotted on the diagram shows the measured values of the coexisting bending moments and axial forces carried by the stiffener and this point is seen to lie reasonably close to the design value.

The method of transverse stiffener design presented herein corresponds to the method for predicting the ultimate shear capacity described earlier. It is semi-empirical and has been shown to lead to a safe design of transverse stiffeners, although it is slightly conservative. It is, however, rather complex for use at the design stage and could well be simplified as a result of further studies.

1.4   LONGITUDINALLY STIFFENED GIRDERS

To obtain further economy and efficiency in the design of plate girders, slender web plates are often reinforced by longitudinal, in addition to transverse, stiffeners. The main function of such longitudinal stiffeners is to increase the buckling resistance of the web; an effective stiffener will remain straight, thereby sub-dividing the web and limiting the buckling to the smaller sub-panels. The resulting increase in ultimate strength can be very significant.

1.4.1   Ultimate Load Capacity of Longitudinally Stiffened Girders
(a) *Girders Loaded Predominantly in Shear*
When the shear failure mechanism, illustrated earlier in Figs. 1.2(c) and 1.3,

was first proposed by Rockey *et al.* (1974), it was also postulated that a longitudinally reinforced plate girder, subjected predominantly to shear, would develop a similar collapse mechanism, provided that the stiffeners were sufficiently rigid to remain effective right up to failure. This observation was based on the study of the failure modes of a number of longitudinally stiffened girders, similar to that shown in Fig. 1.15. It is clear from this photograph that large areas of the sub-panels have remained perfectly flat at failure and that the tension field appears to extend over the complete depth of the web.

On this basis, in order to provide a simple method for design calculations, it may be assumed that the main effect of the longitudinal stiffeners is to increase the buckling resistance of the web. Once one of the sub-panels of the web has buckled, the post-buckling tension field action develops over the whole depth of the web panel and the influence of the stiffeners upon the tension field may be neglected.

Thus, the equations established earlier in Sections 1.2.1 and 1.2.2 to define the shear capacity for the portion SC of the interaction curve plotted in Fig. 1.6 may also be applied to longitudinally stiffened girders. The only difference is that the value of the critical shear stress (τ_{cr} and τ_{crm}) in these equations must be modified. In the case of a longitudinally stiffened web, the value of the critical stress must be calculated for each individual sub-panel. When the stiffened web is subjected to bending moments in addition to shear, the appropriate distribution of axial stresses must be allowed for in the determination of the critical stress. In the case of a transversely stiffened web, the complete web was subjected to a bending stress distribution, as discussed in Section 1.2.2(b), so that the modified shear buckling stress could be determined from eqn (1.14). However, for a

FIG. 1.15.   A typical longitudinally stiffened girder after failing in shear (Rockey *et al.*, 1979).

H. R. EVANS

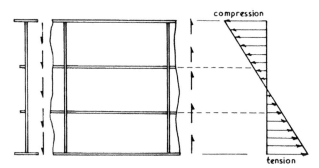

FIG. 1.16.   Stress distribution for a longitudinally stiffened girder under shear and bending.

longitudinally stiffened web such as that shown in Fig. 1.16, the upper sub-panel will be subjected predominantly to longitudinal compression, the distribution of stress across the central panel will vary linearly from compression to tension and the lower sub-panel will be subjected mainly to tension. The appropriate shear buckling stress can be established in each case from elastic buckling theory, conservatively assuming simply supported boundaries for simplification, as discussed in some detail by Ardali (1980).

The buckling resistance of the weakest sub-panel is then substituted into the relevant equations. Also, in the estimation of the direction of the tension field (θ) in eqn (1.8), θ for the longitudinally stiffened web is taken as being approximately equal to two-thirds the inclination of the diagonal of the overall web panel. In this way, the determination of the ultimate load capacity of a longitudinally reinforced girder failing in the shear mode becomes no more difficult than that for a transversely stiffened girder.

(b) Girders Loaded in Pure Bending

When girders are subjected to high bending moments so that collapse occurs in a bending, rather than in a shear mode, i.e. in the region beyond point C of the interaction diagram in Fig. 1.6, the introduction of longitudinal stiffeners can increase the ultimate load capacity significantly. In girders with slender webs, failure of the compression flange will normally be preceded by buckling in the compression zone of the web, as discussed in Section 1.2.3. Correct deployment of the stiffeners in the compression zone can delay, or even eliminate, this buckling, thereby reducing the tendency for any shedding of load from the web to the flange and increasing the failure load.

The ultimate moment capacity (M_u) of a longitudinally stiffened girder can again be determined from the empirical formula proposed by Cooper, as given in eqn (1.18). Indeed, several tests were carried out by Cooper on longitudinally stiffened girders so that the accuracy of the empirical formula is well-established.

(c) Experimental Verification

The application of the shear mechanism approach to longitudinally stiffened girders has again been verified by comparison with available test data and the results of some 30 tests were summarised by Rockey et al. (1978). It was found that the mean value of the ratio between the predicted and measured failure loads for these tests was 0·995, with a standard deviation of 0·071, thus proving the accuracy of the proposed approach.

An example of the results obtained is presented in Fig. 1.17 where the results of four tests carried out by Rockey et al. (1977a) are summarised. As shown by the inset diagrams, the test girders were identical, other than for the number of longitudinal stiffeners employed; test panels varied from an unstiffened case to a panel with 3 longitudinal stiffeners, as shown. The theoretically predicted interaction diagrams are plotted and the four experimental values are superimposed upon them. The excellent agreement

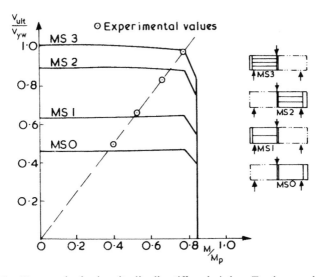

FIG. 1.17.   Test results for longitudinally stiffened girders (Rockey et al., 1977a).

between the predicted and measured values is apparent and the results also show the significant increase in ultimate load capacity achieved by the introduction of the longitudinal stiffeners.

1.4.2 Design of Longitudinal Stiffeners
The two main problems facing the designer of a longitudinally stiffened girder are, firstly, how many stiffeners should be used and how they should be positioned and, secondly, what the dimensions of the stiffeners should be. These two aspects will now be considered individually.

(a) Number and Positioning of Stiffeners
It is difficult for the designer to determine the effectiveness of providing web stiffeners. An increase in ultimate-load capacity can be achieved, either by thickening the web, or by reducing the spacing of the transverse stiffeners, or by providing longitudinal stiffeners. Although the introduction of stiffeners is normally effective in producing a structure of high strength/weight ratio, the additional fabrication costs involved do not always make such solutions cost-effective.

To provide some guidance, a parametric study employing the tension field mechanism was carried out by Ardali (1980) in which some 15 000 girders were analysed. The girders were subjected to various combinations of shear and bending although the study was restricted to failure in the shear mode. The effects of varying the web slenderness ratio (d/t), the web aspect ratio (b/d) and the flange strength (M_p^*) were studied with the range of variation of each parameter extending well beyond those values normally encountered in practice. In each case, the effect of introducing up to eight longitudinal stiffeners was considered.

In the case of pure shear, the stiffeners were spaced equally over the depth so as to divide the web into equal sub-panels. This is the optimum spacing for the shear case, since the shear-buckling capacity of each of the sub-panels then becomes equal. When bending moments were applied to the web together with the shear, the stiffener spacing was adjusted so that the panels subjected to the highest compressive stresses were smaller than the others.

A typical set of curves that was obtained for a high shear loading case is presented in Fig. 1.18. The percentage increase in ultimate-load capacity produced by stiffening is plotted against the slenderness ratio (d/t) of the web. A family of curves is presented, each curve corresponding to a different number of longitudinal stiffeners.

Although pertaining to a particular web aspect ratio and flange strength,

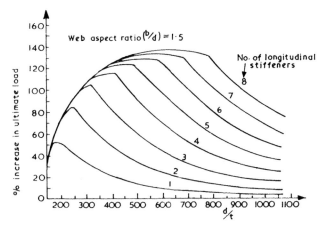

FIG. 1.18.   Percentage increase in strength of shear webs with number of
longitudinal stiffeners (Ardali, 1980).

these curves are representative of all the curves obtained in the study. It is
observed that all the curves lie within an envelope which has a steep slope at
low values of d/t but levels off as the slenderness increases. It is also noted
that the curve corresponding to any specific number of stiffeners branches
down from the envelope at a certain slenderness ratio. Until this value of d/t
is reached, no benefit is gained by using more stiffeners than the optimum
number. For more slender webs, further stiffeners must be added to
improve the load-carrying capacity. Thus, such curves enable the designer
to choose the optimum number of stiffeners for any web slenderness.

In the case of girders subjected to pure bending, several studies based on
linear buckling theories, have been carried out to determine the optimum
stiffener positioning. The results of such studies were summarised in the
work of Owen et al. (1970), where full references to the earlier investigations
were given.

(b) Stiffener Dimensions
In its prediction of the ultimate load, the tension field mechanism solution
assumes that the longitudinal stiffeners remain fully effective right up to
failure. However, it gives no direct information as to what the dimensions
of the stiffeners should be in order to ensure that this requirement is
satisfied.

As in the case of the transverse stiffeners discussed in Section 1.3, linear
buckling theories can be employed to establish the minimum value of

stiffener rigidity (γ^*) required to ensure that the longitudinal stiffeners remain straight at first buckling, thereby limiting buckling to the individual sub-panels of the web. Such optimum values of γ^* can be established for various combinations of bending and shear loads and various stiffener positions. Full references to such studies may again be found in the work of Owen et al. (1970).

However, these investigators and several others, notably Massonnet (1960) and Meszaros and Djubek (1966), have shown that this value of stiffener rigidity must be increased by a factor ranging from 3 to 7 to ensure that the stiffener remains straight right up to the failure load. The actual value of the factor to be taken in any particular case depends on several parameters, particularly on the ratio between the applied bending moment and shearing force and upon the number and positioning of the stiffeners.

One disadvantage of such an empirical approach is that many graphs are required to cover every possible case that may be encountered in design. Indeed the new British Code of Practice for the Design of Steel Bridges, BS 5400 Part 3 (1982), departs from this philosophy and recommends that longitudinal stiffeners should be designed as struts following a procedure rather similar to that described in Section 1.3 for transverse stiffeners. This procedure was discussed in some detail by Chatterjee (1981).

Few experimental studies of the behaviour of longitudinal stiffeners on webs subjected primarily to shear have been reported. Rockey et al. (1979) described a series of six tests on such girders. All the girders tested were identical, other than for the rigidity of the longitudinal stiffeners, and the characteristics of the girders are shown on the inset diagram in Fig. 1.19.

The experimental results are also summarised in Fig. 1.19 where the measured failure load is plotted against the stiffener rigidity. The experimental points are identified by the labels ascribed to the various girders and, as indicated, girder SD3 was unstiffened, SD8 had very light stiffeners ($\gamma = 2\cdot63$) and the stiffener rigidity was then increased progressively up to a γ-value of 216·4 for girder SD4.

The heavy stiffeners employed on girders SD4, SD5 and SD6 proved fully adequate during the tests and little variation is observed in their ultimate capacity. The very light stiffeners on girder SD8 did fail quite clearly and the plotted values show that the full load-carrying capacity of this girder was not achieved.

The remaining girder SD7 did achieve its full capacity but signs were observed during the test that the stiffeners were close to failure. This particular stiffener rigidity of $\gamma = 13\cdot5$ must, therefore, have been close to the optimum for a girder of these dimensions when subjected to shear

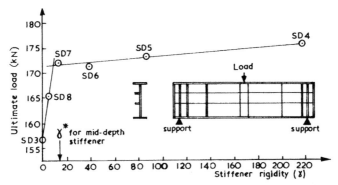

FIG. 1.19.   Variation of ultimate load with longitudinal stiffener rigidity (Rockey *et al.*, 1979).

loading. This rigidity value is, in fact, very close to the γ^* value obtained from linear buckling theory for a single stiffener at mid-depth, as indicated on the horizontal axis of Fig. 1.19.

Although these experimental results are only strictly valid for the particular girder dimensions considered, further experimental evidence was obtained by Rockey *et al.* (1977*b*). In their tests on large-scale box girders in shear, these authors measured the strains developed in the longitudinal web stiffeners and observed that these strains remained low in the post-buckling phase but increased rapidly as the girder failed.

This would suggest that for shear webs, the use of the γ^* concept for the design of longitudinal stiffeners could be satisfactory. However, further experimental evidence, which the author is currently endeavouring to establish, is required to support this view.

1.5   CONCLUSION

This chapter has considered the post-buckling behaviour of plate girders in some detail. Fundamental theoretical relationships have been established and their application in design has been considered. In some instances, semi-empirical relationships have been presented to simplify the design procedure.

In general, the post-buckling behaviour is well understood. However, there are areas, particularly with regard to the behaviour of longitudinal web stiffeners, where further information is required. Such further studies may also lead to simplification of the design procedure for transverse stiffeners.

REFERENCES

ALLEN, H. G. and BULSON, P. S. (1980) *Background to Buckling*, McGraw-Hill, Maidenhead, Berks.

ARDALI, S. (1980) A parametric study of the effect of longitudinal web stiffeners on the behaviour of plate girders. MSc Thesis, University College, Cardiff.

BASLER, K. (1960) Strength of plate girders in shear. Fritz Engineering Laboratory Report No. 231-20, Lehigh University, Bethlehem.

BJORHOVDE, R. (1980) Research needs in stability of metal structures. *Journal of the Structural Division, ASCE*, ST 12, 2425–41.

BRITISH STANDARDS INSTITUTION (1982) Code of Practice for Design of Steel Bridges, BS 5400, Part 3.

CALLADINE, C. R. (1973) A plastic theory for collapse of plate girders under combined shearing force and bending moment. *Structural Engineer*, **51**, 147–54.

CHATTERJEE, S. (1981) Design of webs and stiffeners in plate and box girders. *The Design of Steel Bridges*, Granada Publishing, St. Albans.

CHERN, C. and OSTAPENKO, A. (1969) Ultimate strength of plate girders under shear. Fritz Engineering Laboratory Report No. 328-7, Lehigh University, Bethlehem.

COOPER, P. B. (1965) Bending and shear strength of longitudinally stiffened plate girders. Fritz Engineering Laboratory Report No. 304-6, Lehigh University, Bethlehem.

COOPER, P. B. (1971) The ultimate bending moment for plate girders. *Proceedings of IABSE Colloquium*, London, 291–7.

EVANS, H. R., PORTER, D. M. and ROCKEY, K. C. (1976) A parametric study of the collapse behaviour of plate girders. University College, Cardiff, Report.

EVANS, H. R., PORTER, D. M. and ROCKEY, K. C. (1978) The collapse behaviour of plate girders subjected to shear and bending. *IABSE Proceedings* P-18/78 (Nov.), 1–20.

EVANS, H. R. and TANG, K. H. (1981) A report on five tests carried out on a large-scale transversely stiffened plate girder—TRV3. University College, Cardiff, Report No. DT/SC/8.

FUJII, T. (1968) On an improved theory for Dr Basler's theory. *Proc. 8th IABSE Congress*, New York, 477–87.

HORNE, M. R. (1979) *Plastic Theory of Structures*, Pergamon Press, Oxford.

KOMATSU, S. (1971) Ultimate strength of stiffened plate girders subjected to shear. *Proceedings of IABSE Colloquium*, London, 49–65.

KUHN, P. and PETERSON, J. P. (1947) Strength analysis of stiffened beam webs. *NACA*, TN 1364.

KUHN, P., PETERSON, J. P. and LEVIN, C. R. (1962) A summary of diagonal tension, parts I and II. *NACA*, TN 2661 and 2662.

MASSONNET, C. (1960) Stability considerations in the design of steel plate girders. *Journal of the Structural Division, ASCE*, No. 2350 (ST1), 71–97.

MELE, M. and PUTHALI, R. (1979) Ottimizzazione degli irrigidimenti di travi a parete piena sottile. Relazione Finale, University of Rome, Report CECA R/136.

MESZAROS, I. and DJUBEK, J. (1966) Vplyv Tuhosti Vystuh na Deformativ nost Stien. *Stavebnicky Casopis*, SAV X1V3, Bratislava.

OWEN, D. R. J., ŠKALOUD, M. and ROCKEY, K. C. (1970) Ultimate load behaviour of longitudinally reinforced web plates subjected to pure bending, *IABSE Memoires*, Zurich, 113–48.

ROCKEY, K. C., EVANS, H. R. and PORTER, D. M. (1978) A design method for predicting the collapse behaviour of plate girders. *Proc. Instn. Civ. Engrs.*, Part 2, 85–112.

ROCKEY, K. C., EVANS, H. R. and PORTER, D. M. (1977a) Tests on longitudinally reinforced plate girders subjected to shear. *Proceedings of Conference on Stability of Steel Structures*, Liege.

ROCKEY, K. C., EVANS, H. R. and PORTER, D. M. (1977b) Tests on large scale steel box girders. University College, Cardiff, Report.

ROCKEY, K. C., EVANS, H. R. and TANG, K. H. (1979) An investigation of the rigidity of longitudinal web stiffeners for plate girders. University College, Cardiff, Report.

ROCKEY, K. C., PORTER, D. M. and EVANS, H. R. (1974) The ultimate shear load behaviour of longitudinally reinforced plate girders. *Proceedings of Symposium on Structural Analysis, Non-Linear Behaviour and Techniques, TRRL*, 163–74.

ROCKEY, K. C. and ŠKALOUD, M. (1968) Influence of flange stiffeners upon the load carrying capacity of webs in shear. *Proc. 8th Congress IABSE*, New York, 429–39.

ROCKEY, K. C., VALTINAT, G. and TANG, K. H. (1981) The design of transverse stiffeners on webs loaded in shear—an ultimate load approach. *Proc. Instn. Civ. Engrs.*, Part 2 (Dec.), 1069–99.

WAGNER, H. (1929) Ebene blechwandtrager mit sehr dunnen Stegblechen. *Z. Flugtech Motor Luftsch*, **20**.

WINTER, G. (1947) Strength of thin steel compression flange. Cornell University Engineering Experiment Station Bulletin, 35/3.

Chapter 2

ULTIMATE SHEAR CAPACITY OF PLATE GIRDERS WITH OPENINGS IN WEBS

R. Narayanan

Department of Civil and Structural Engineering,
University College, Cardiff, UK

SUMMARY

Theoretical methods of computing the ultimate shear capacity of plate girders containing circular or rectangular web holes are described. The reinforcement requirements to restore the strength lost by the introduction of the holes are discussed. The theory proposed is based on the consideration of the equilibrium of the girders in the collapse state. The ultimate load is computed as the sum of four contributions, viz. (i) the elastic critical load, (ii) the load carried by the membrane tension in the post-critical stage, (iii) the load carried by the flange and (iv) the load carried by the reinforcement, if any. The results of some 70 ultimate load tests on plate girders containing various forms of web holes are summarised. The strength predictions obtained by using the theory proposed are shown to be sufficiently accurate in comparison with the values observed in test girders.

NOTATION

b	Clear width of web plate between vertical stiffeners
b_e	Effective width of membrane field
b_f	Breadth of flange (in compression and in tension)
b_o	Width of the rectangular opening in web plate

c	Distance between hinges (in compression and tension flange)
c_r	Distance between the hinges on the reinforcement
D	Diameter of the circular opening in web plate
d_o	Depth of the rectangular opening in web plate
d	Clear depth of web plate between flanges
l	Overhanging length of the reinforcement
t	Thickness of web plate
t_f	Thickness of flange plate (in compression and in tension)
t_r	Thickness of reinforcing strip
w_r	Width of reinforcement

E	Modulus of elasticity
M_p	Plastic moment of resistance of flange plates $(b_f t_f^2/4)\sigma_{yf}$
M_{pr}	Plastic moment of resistance of reinforcement
V	Applied shear load $= W/2$
V_{ult}	Ultimate shear force

α	Angle of inclination of the diagonal of the rectangular cutout $=$ arc tan (d_o/b_o)
β	Non-dimensional coefficient for evaluating the overhang of reinforcing strip
δ	Diameter of an imaginary circle to replace a rectangular cutout in the membrane field
θ	Angle of inclination of the tensile membrane stress σ_t^y
θ_d	Angle of inclination of the panel diagonal $=$ arc tan (d/b)
θ_m	Optimum value of θ
μ	Poisson's ratio
κ_0	Non-dimensional buckling coefficient for unperforated webs
κ	Non-dimensional buckling coefficient for perforated webs
τ_{cr}	Critical shear stress in the web
$(\tau_{cr})_{red}$	Reduced value of critical shear stress
$(\tau_{cr})_{mod}$	Modified value of critical shear stress due to the presence of the reinforcement
σ_{yf}	Yield stress of flange members
σ_t^y	Membrane stress in the post-critical stage
σ_{yw}	Yield stress in web plate
σ_{yr}	Yield stress of reinforcement

Note: other symbols are defined in the paper as they appear.

2.1 INTRODUCTION

Designers frequently find it necessary to introduce openings in the webs of plate and box girders in order to provide services and for affording inspection. The introduction of an opening alters the stress distribution within the member and will, in most cases, influence its collapse behaviour.

In general, webs in built-up girders are made up of relatively thin plates, having web-slenderness values, (d/t), in excess of 200; they are subjected predominantly to shear loading and their collapse loads in shear are significantly higher than their respective elastic critical loads. Even though satisfactory methods of determining the ultimate shear capacity of plate girders having slender webs have been proposed by several researchers (European Convention for Constructional Steelworks (ECCS), 1976; Narayanan and Adorisio, 1983), very little information is available on the behaviour of these structures containing holes.

The available literature concentrates largely on moderately thick webs having web-slenderness (d/t) values in the region of 50–80. The available methods of analysis for such webs have been reviewed by Redwood in an earlier volume in this series (Redwood, 1983). Mathematical models for the design of thick webs containing perforations have been developed and are based on plastic analysis of structures (Redwood, 1972; Redwood and Shrivastava, 1980). However, these methods are not valid for *thin* webs of the type used in plate and box girders on account of the high web-slenderness values associated with such structures. Typical values of slenderness for these webs lie in the region 200–350 and they invariably buckle prior to the actual collapse of the girder. A knowledge of the effects of web openings on the web instability and its collapse behaviour is of vital importance to the designers of these structures.

The first published paper of relevance to thin webs containing various types of openings was by Hoglund (1971) who reported on statically loaded plate girders containing circular and rectangular holes and subjected to transverse loading. He carried out some 12 tests on four simply supported plate girders, having both circular and rectangular web holes. The web plates had slenderness values in the region 200–300 and were not stiffened, in keeping with Swedish practice. The web holes were located both in high shear and in high moment zones. The tests revealed that the girders having holes in the high shear zone failed at loads significantly lower than those which had holes in the high moment zones. His experiments have, therefore, indicated the relative importance of shear failure criteria in plate girders with perforated webs. Based on the somewhat limited experimental data obtained by him, he proposed a theoretical model for the evaluation of the ultimate capacity of girders containing holes. The method proposed by him consisted of tension fields with stress equal to the tensile yield stress and compression fields with stress levels estimated (somewhat empirically) from his experimental observations.

The only other paper of relevance to plate girders containing openings in webs is by Narayanan and Rockey (1981) who presented the results of some 20 tests and proposed an approximate method of analysis of these girders; this is discussed more fully in Section 2.3.

More accurate analytical models are developed in this chapter and are based on the theoretical method for estimating the ultimate capacity of plate girders suggested by Porter *et al.* (1975). The values of ultimate shear computed by the analytical models suggested in this chapter are compared with the measured collapse loads obtained in a comprehensive experimental programme of tests covering parametric variations in web slenderness, hole shape, size and location and the aspect ratio of the panel. The methods proposed are shown to be sufficiently accurate for design purposes.

2.2   OBSERVED FAILURE PATTERN IN PLATE GIRDERS

The simple case of a web bounded by two flanges and three vertical stiffeners will now be considered (Fig. 2.1(a)). To begin with, the shear is resisted by the web plate up to the elastic critical load. Any further increase of load does *not* cause a rapid collapse of the girder but results in the formation of buckles in a waveform parallel to the tensile direction. A small band of web plate along the tensile diagonal commences to behave in a manner similar to a tension member of a corresponding truss (Fig. 2.1(b)); the development of membrane tension in the web is denoted as 'tension field action' and enables the web to sustain loads well in excess of the elastic critical load. A consequence of the membrane tension in the web is the inward pulling of the flanges, under increasing loads. Eventually, plastic hinges are formed in the flanges, and trigger the collapse of the girder (Fig. 2.1(c)). To sum up, the collapse of the plate girder is accounted for by three contributions:

1.   the elastic critical load;
2.   the tension field;
3.   the contribution of the flanges.

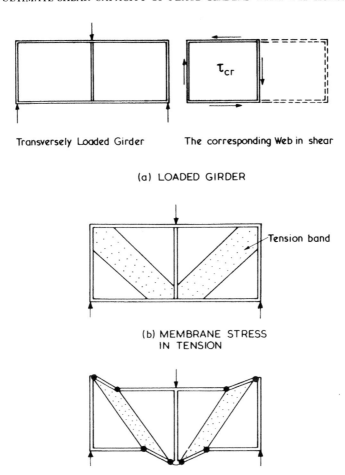

(a) LOADED GIRDER

(b) MEMBRANE STRESS
IN TENSION

(c) FORMATION OF HINGES
IN FLANGES

FIG. 2.1.   Failure pattern of a plate girder.

A fuller description of the collapse behaviour of plate girders is given by Evans in Chapter 1 of this book. Porter *et al.* (1975) have proposed an equilibrium solution for evaluating the ultimate shear (V_s) of a plate girder with slender webs using the above concepts. The ultimate shear is given by eqn (1.4) in Chapter 1:

$$V_s = \tau_{cr}dt + \sigma_t^y t \sin^2 \theta (d \cot \theta - b + c) + \frac{4M_{pf}}{c} \qquad (2.1)$$

where τ_{cr} = elastic critical shear stress for the web, σ_t^y = membrane stress in the web in the post-critical stage, b = width of the web panel, c = distance between the hinge formed in the flange, d = depth of the web, t = thickness of the web, θ = angle of inclination of the tensile membrane stress σ_t^y and M_{pf} = fully plastic moment of the flange.

The values of c and σ_t^y in the above equation can be obtained in terms of known quantities (see below); however, θ is unknown. As this is an 'equilibrium' solution, θ is evaluated by trial and error to obtain a *maximum* value for V_s.

By considering the equilibrium of the flange at the instant of collapse, the value of c can be evaluated (see eqn (1.5), Chapter 1):

$$c = \frac{2}{\sin \theta} \sqrt{\left(\frac{M_{pf}}{\sigma_t^y t}\right)} \tag{2.2}$$

Membrane stress σ_t^y can be calculated by applying Von Mises criterion, if the yield stress σ_{yw} and the elastic critical stress τ_{cr} of the web are known:

$$\sigma_t^y = -\tfrac{3}{2}\tau_{cr} \sin 2\theta + \sqrt{(\sigma_{yw}^2 + \tau_{cr}^2 \{(\tfrac{3}{2}\sin 2\theta)^2 - 3\})} \tag{2.3}$$

The elastic critical stress in shear, τ_{cr}, for a rectangular web is given by (Timoshenko and Gere, 1961):

$$\tau_{cr} = \kappa_0 \frac{\pi^2 E}{12(1-v^2)} \left(\frac{t}{d}\right)^2 \tag{2.4}$$

where κ_0 is a non-dimensional coefficient, E is the Young's modulus and v is the Poisson's ratio.

The value of κ_0 for a web simply supported along the edges is given by

$$\kappa_0 = 5 \cdot 35 + 4 \left(\frac{d}{b}\right)^2 \qquad \text{when } \frac{b}{d} > 1 \cdot 0 \tag{2.4a}$$

2.3   WEBS CONTAINING CENTRAL CIRCULAR HOLES

A systematic study of thin webs (i.e. with values of web slenderness (d/t) in the range 200–350) containing central circular and rectangular openings has been in progress at the University College, Cardiff since 1977. Narayanan and Rockey (1981) reported on the ultimate load tests carried out by them on 20 panels containing centrally placed circular holes and subjected to shear loading. The parameters varied included web slenderness, flange stiffness and hole size. The tests showed that the ultimate capacity of webs dropped almost linearly with the increase in the diameter of the cutout. The failure mechanism observed by them in plate girders with web holes was similar in form to that observed in girders with unperforated webs (see Chapter 1 of this book), the only difference being in the position of hinges. They proposed an approximate method to assess the ultimate capacity of a plate girder with the perforated web. The method consisted of linearly interpolating between the value of V_s for an unperforated web obtained from eqn (2.1) and the Vierendeel load, V_v obtained as described below.

If the diameter of the hole, D, covers the full depth, d, of the girder, the failure would be essentially due to Vierendeel mechanism with hinges formed centrally at the top and bottom flanges (see Fig. 2.2). The corresponding collapse load, V_v, is given by

$$V_v = \frac{8M_{pf}}{b} \tag{2.5}$$

For a hole diameter smaller than D, V_{ult}, the ultimate shear capacity can be approximately evaluated by linear interpolation between the values of V_v and V_s:

$$V_{ult} = V_v + \left(\frac{V_s - V_v}{d}\right)(d - D) \tag{2.6}$$

Even though the above method is somewhat empirical, it has been shown that the predictions of ultimate shear capacity obtained, are close to the experimentally observed values (Narayanan and Rockey, 1981).

The proposed equilibrium solution also consists of evaluating the strength of the girder as the sum of three contributions as before, viz:

1.   the critical load on the web;
2.   the load carried by the membrane tension field;
3.   the load carried by the flange when the collapse is imminent.

When the shear on the plate girder is increased the web plate reaches its elastic critical value first. Rockey et al. (1967) investigated the buckling of a *square* plate under shear loading using a finite element method. They showed that as a result of the introduction of a central circular hole, the buckling resistance decreased, the amount depending upon the ratio of the hole size to the width of (the square) plate. They also showed that the reduced value of the elastic critical stress in shear $(\tau_{cr})_{red}$ could be obtained by an equation similar to eqn (2.4); the only change is that a reduced value

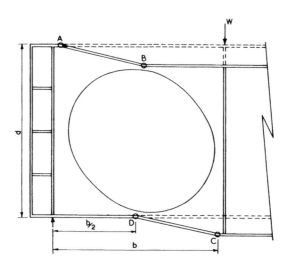

Note: Two hinges are formed at the centre of the
flanges when the diameter of the hole is
nearly equal to the depth

FIG. 2.2.   Vierendeel action.

of the buckling coefficient κ (appropriate to the perforated web) should be
used.

Using a finite element method, Narayanan and Der Avanessian (1982a)
studied the buckling behaviour of square and rectangular plates for two
support conditions, viz.

i.   When all the edges of the plate were simply supported
ii.   When all the edges were fixed.

Figure 2.3 summarises the relationship obtained in that study. They
concluded that the relationship between the buckling coefficient κ_0,
appropriate to a perforated web, and the diameter of the hole was nearly
linear.

$$(\tau_{cr})_{red} = \kappa \frac{\pi^2 E}{12(1 - v^2)} \left(\frac{t}{d}\right)^2 \tag{2.7}$$

where

$$\kappa = \kappa_0 \left(1 - \frac{D}{d}\right) \tag{2.8}$$

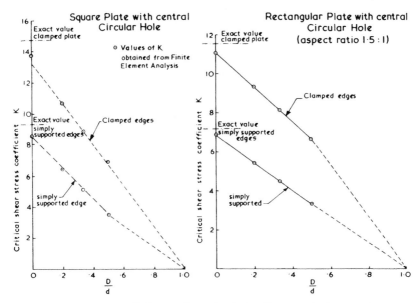

FIG. 2.3.   Buckling coefficients when the hole diameter is increased. (From Narayanan and Der Avanessian, 1982a.)

For simply supported edges, κ_0 is evaluated from eqns (2.4a). For edges which are fixed κ_0 is evaluated from:

$$\kappa_0 = 8.98 + 5.6 \left(\frac{d}{b}\right)^2 \qquad \text{for } \left(\frac{b}{d} > 1.0\right) \tag{2.8a}$$

The reduced value of the elastic critical stress can, therefore, be obtained from the approximate relation:

$$(\tau_{cr})_{red} = \kappa_0 \left(1 - \frac{D}{d}\right) \frac{\pi^2 E}{12(1 - v^2)} \left(\frac{t}{d}\right)^2 \tag{2.9}$$

The web is, of course, capable of sustaining further loading in the post-critical stage; the additional load is carried by the membrane stresses (σ_t^y) developed in the web. These form two bands one above and the other below the cutout (see Fig. 2.4). This phenomenon has been observed in the pattern of buckles seen in the girders tested by Narayanan and Rockey (1981).

A part of the load is carried by the flanges when collapse is about to occur; the moment capacity of the flange is equal to its plastic moment

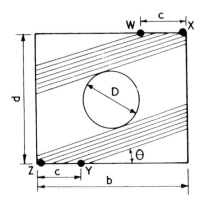

FIG. 2.4.   Tension bands formed for small circular holes.

(M_{pf}). The ultimate capacity of the girder is obtained by adding the contribution due to the flange stiffness to the load taken by the web.

Figure 2.5 shows the position of hinges (W, X, Y, Z) at the instant of failure. Let us consider the part of the panel to the right of WY. Across the diameter AB no forces can act consequent on the stresses being zero around the periphery of the hole. Let the two tensile membrane forces on each of the two tension bands (one above and the other below the hole) be F_s. The

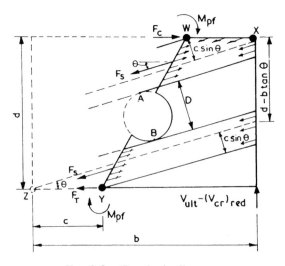

FIG. 2.5.   Free body diagram.

vertical components of these two forces add up to $2F_s \sin \theta$. From the geometry of the structure,

$$2F_s = \sigma_t^y t[2c \sin \theta + (d - b \tan \theta) \cos \theta - D] \tag{2.10}$$

The internal plastic hinge will form at the position of maximum bending moment, where the shear force is zero. The hinge distance (c) can therefore be evaluated as before and is given by eqn (2.2). Membrane stress σ_t^y can be evaluated by applying the Von Mises criterion (eqn 2.3), remembering that the critical stress is the reduced value $(\tau_{cr})_{red}$ calculated from eqn (2.9).

$$\sigma_t^y = -\tfrac{3}{2}(\tau_{cr})_{red} \sin 2\theta + \sqrt{(\sigma_{yw}^2 + (\tau_{cr})_{red}^2[(\tfrac{3}{2} \sin 2\theta)^2 - 3])} \tag{2.11}$$

Narayanan and Der Avanessian (1981a) have demonstrated that in calculating $(\tau_{cr})_{red}$, the value of κ appropriate to a web fixed at its edges should be used. The reason for this is as follows: the relative stiffness of the flange in comparison with the web increases significantly when the hole is introduced in the web and the behaviour of the web plate is closer to one having encastré supports.

From a consideration of vertical equilibrium of forces, V_{ult}, the ultimate shear capacity can be evaluated:

$$V_{ult} = (V_{cr})_{red} + 2F_s \sin \theta \tag{2.12}$$

Substituting for F_s, and simplifying, we obtain

$$V_{ult} = 2c\sigma_t^y t \sin^2 \theta + \sigma_t^y td(\cot \theta - \cot \theta_d) \sin^2 \theta - \sigma_t^y tD \sin \theta + (\tau_{cr})_{red}dt \tag{2.13}$$

This equation is valid for all holes having $D \le d\cos \theta - b \sin \theta$.

The above limitation is, however, not highly restrictive, as it includes holes of all practical proportions. Larger cutouts are unlikely to be met in practice; however, a method of solution for girders with large holes has been suggested by Narayanan and Der Avanessian (1981a).

It will be seen that when the diameter D, is zero, eqn (2.13) reduces to the solution for a girder without a web hole. For given dimensions of a girder and for a specified value of hole diameter, c and σ_t^y can be evaluated in terms of θ, which is the unknown angle of inclination of the membrane stress. Since eqn (2.13) has been obtained as an 'equilibrium' solution, the optimum angle of θ will produce a maximum value for V_{ult}. It is, therefore, possible to obtain the maximum value for V_{ult} by choosing several trial values of θ. A parametric study of plate girders containing various hole diameters shows that for girders representative of normal sizes, the optimum angle drops linearly with an increase in the value of D. A typical variation in θ_m, the optimum value of θ, is shown in Fig. 2.6.

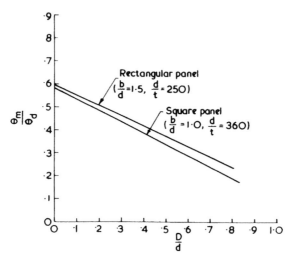

FIG. 2.6.   Variation of optimum angle θ_m with the diameter of cutout.

2.4   WEBS CONTAINING ECCENTRICALLY PLACED CIRCULAR HOLES

Since the tension field in an unperforated web is developed predominantly along a diagonal band, it is wise to locate openings away from this band, so that the girder does not suffer any significant drop in strength. Experiments carried out by the author confirm that by locating the holes away from the tension band, it is possible to obtain a larger ultimate capacity compared with girders having centrally located web holes (Narayanan, 1980).

When the diameter of the hole is small and its location is such that it does not intersect the idealised tension band (see Fig. 2.7), the ultimate shear capacity can be computed by omitting the term containing D in eqn (2.13):

$$V_{ult} = 2c\sigma_t^y t \sin^2 \theta + \sigma_t^y td(\cot \theta - \cot \theta_d)\sin^2 \theta + (\tau_{cr})_{red}dt \quad (2.14)$$

Quantities c and σ_t^y are given by eqns (2.2) and (2.11) respectively.

Equation (2.14) is valid so long as the spread of the tension field in the tension flange, g, is greater than the hinge distance, c (see Fig. 2.7). In the above equation $(\tau_{cr})_{red}$ refers to the elastic critical stress for the webs with eccentrically located holes. Narayanan and Der Avanessian (1982a) have computed the values of $(\tau_{cr})_{red}$ for several cases of plates with eccentrically located holes using a finite element method; Fig. 2.8 gives a typical relation

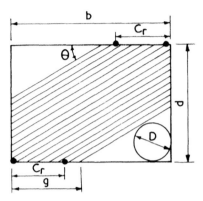

FIG. 2.7.   Small diameter cutout.

between κ and the distance of the centre of the hole to the centre of panel (for the case of $D/d = 0.33$). It is seen that the value of the elastic critical stress improved when the centre of the hole moved away from the panel centre and away from the tension field. The increase in the value of κ depended on panel dimensions, hole size and edge support conditions. Values of $(\tau_{cr})_{red}$ obtained from eqn (2.9) are therefore smaller than the correct values appropriate to these holes. However, designers will find eqn (2.9) simple to use, and the errors introduced in the ultimate load prediction are small and conservative. This is because the critical load forms only a small part of V_{ult}; hence it is proposed to use the approximate relationship given by eqn (2.9) for evaluating V_{ult} in eqn (2.14). As before the maximum value of V_{ult} is obtained by trial and error.

When the diameter of the opening is large (Fig. 2.9) the hinge distances in the compression flange (c_c) and in the tension flange (c_t) are not equal. The spread of the tension field is obtained from

$$g = b - \frac{D}{2}\left(\cot\frac{\theta}{2} + 1\right) \tag{2.15}$$

Experiments on plate girders with large holes show that the hinge in the tension flange is formed below the vertical diameter of the hole as shown:

$$c_t = b - \frac{D}{2} \tag{2.16}$$

There is no change in the location of the hinges in the compression flange, i.e. $c_c = c$, calculated as before.

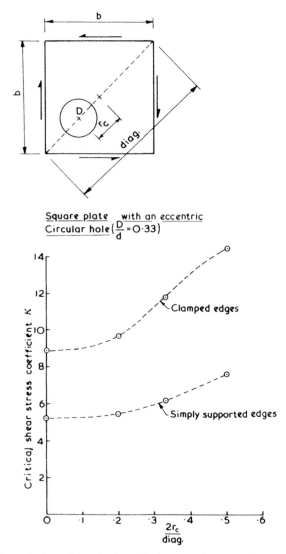

FIG. 2.8.   The variation of the elastic critical stress with the hole location. (From
Narayanan and Der Avanessian, 1982a.)

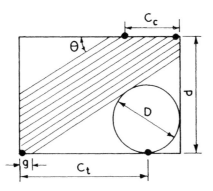

FIG. 2.9.   Large diameter cutout.

From Fig. 2.10(a), the vertical component of force on the side BC is given by

$$F_{BC}^v = \sigma_t^y t \left(d \cot \theta - \frac{D}{2} \left(\cot \frac{\theta}{2} + 1 \right) \right) \sin^2 \theta \qquad (2.17)$$

(Note: the superscript v on F_{BC}^v represents the vertical component of the force.)

Similarly the vertical components of forces acting on the part of the flanges between the hinges (AB and DC) are given by

$$F_{AB}^v = \sigma_t^y t c_c \sin^2 \theta$$

$$F_{DC}^v = \sigma_t^y t \left(b - \frac{D}{2} \left(\cot \frac{\theta}{2} + 1 \right) \right) \sin^2 \theta$$

When the sway mechanism shown in Fig. 2.10(b) occurs at failure, the ultimate shear can be computed by considering the virtual work done by the rotations of ϕ_1 and ϕ_2 at the hinges:

$$\left\{ \begin{matrix} \text{Work done by} \\ (V_{ult} - V_{cr}) \end{matrix} \right\} = \left\{ \begin{matrix} \text{Work done by} \\ F_{AB}^v \end{matrix} \right\} - \left\{ \begin{matrix} \text{Work done by} \\ F_{DC}^v \end{matrix} \right\}$$

$$+ \left\{ \begin{matrix} \text{Work done on} \\ \text{face BC} \end{matrix} \right\} + \left\{ \begin{matrix} \text{Rotations of} \\ \text{AB and DC} \end{matrix} \right\}$$

$$(V_{ult} - V_{cr}) c_c \phi_1 = F_{AB}^v (\tfrac{1}{2} c_c \phi_1) - F_{DC}^v (\tfrac{1}{2} g \phi_2)$$

$$+ F_{BC}^v (c_c \phi_1) + 2 M_{pf} (\phi_1 + \phi_2)$$

$$c_c \phi_1 = c_t \phi_2$$

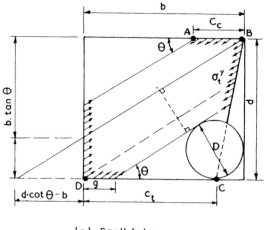

(a) Small hole

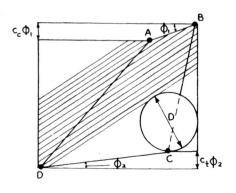

(b) Large hole

FIG. 2.10. Formation of mechanisms with an eccentrically located hole.

Substituting for F_{AB}^v, F_{DC}^v and F_{BC}^v, and after simplification, we obtain

$$V_{ult} = 0 \cdot 5\sigma_t^y t c_c \sin^2 \theta - 0 \cdot 5\sigma_t^y t \left(b - \frac{D}{2} \left(\cot \frac{\theta}{2} + 1 \right) \right) \sin^2 \theta$$

$$+ \sigma_t^y t \left(d \cot \theta - \frac{D}{2} \left(\cot \frac{\theta}{2} + 1 \right) \right) \sin^2 \theta + 2M_{pf} \frac{c_c + c_t}{c_c c_t} + (\tau_{cr})_{red} dt$$

$$(2.18)$$

This equation is valid for the case of $g < c_c$.

For a given size of plate girder, the only unknown in eqn (2.18) is θ. As before the maximum value of V_{ult} can be obtained by trial and error to yield an optimum value θ_m for θ.

Figure 2.11 shows the variation of the angle θ_m with the increase in diameter of the cutout for a girder having $d/t = 360$; it can be seen that θ_m is constant for diameters up to about $0.3d$. Figure 2.12 shows a theoretical relationship between c_c/b and D/d. For diameters up to $0.3d$, $c_c = c_t$. For a

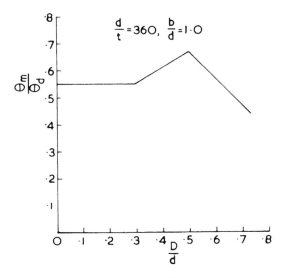

FIG. 2.11.   Variation of optimum angle with the increase in diameter of eccentric cutout (square panel).

range of d/t values, it has been found that θ_m remains constant and $c_c = c_t$ so long as the diameter of the hole does not exceed about 0.3 times the depth of the girder; for this range of diameters eqn (2.14) governs; thereafter eqn (2.18) must be used.

A parametric study of the influence of hole diameters on the ultimate capacity indicates that small diameter eccentrically located holes (shown in Fig. 2.7) having $D \not> 0.25d$ do not cause a significant drop in the strength of the girders; when the holes are larger, the ultimate strength drops dramatically.

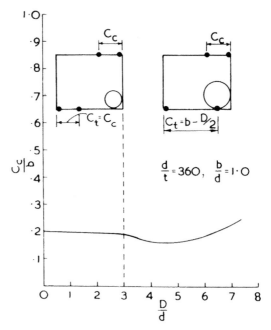

FIG. 2.12. Variation of hinge distance with the diameter of eccentric cutout (square panel).

2.5 WEBS CONTAINING CENTRALLY LOCATED RECTANGULAR HOLES

The strength of the plate girder with a central rectangular opening can be calculated in a manner similar to the above, by summing up the three contributions, viz. the critical load on the web, the load carried by the membrane stress in the post-critical stage and the load carried by the flanges. Two tension bands, one above and the other below the web-hole (analogous to those developed for circular holes) are proposed for the computation of the strength of the girder (see Fig. 2.13).

Shanmugam and Narayanan (1982) and Narayanan and Der Avanessian (1982a) have used finite element methods to evaluate the elastic critical stress of square plates containing square and rectangular holes; Fig. 2.14 shows the results of the latter study. Based on these studies, the following approximate formula for the elastic critical shear stress is

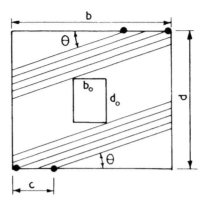

FIG. 2.13.   Tension bands in rectangular openings.

suggested. As the critical load forms only a small part of the ultimate load, the error introduced by this approximate formula is very small.

$$(\tau_{cr})_{red} = \frac{\kappa_0 \pi^2 E}{12(1 - v^2)} \left(\frac{t}{d}\right)^2 \left[1 - \alpha_r \sqrt{\frac{A_c}{A}}\right] \qquad (2.19)$$

where A = total area of the plate (including the cutout), A_c = area of the cutout, κ_0 = coefficient for shear buckling stress appropriate to an unperforated web encastré at its edges (given by eqn (2.8a), α_r = a coefficient, depending on the end conditions and has a value of $1 \cdot 25$ for clamped edges.

Four hinges are formed at the instant of collapse in a manner similar to the girders with circular holes. These are located at W, X, Y and Z in Fig. 2.15, which shows the free body diagram to the right of WY. Across AB, there are no forces and the length δ is equivalent to the diameter of an imaginary central circular cutout

$$\delta = \sqrt{(b_o^2 + d_o^2)} \sin(\alpha + \theta) \qquad (2.20)$$

where α is the angle of inclination of the diagonal of the cutout, i.e. $\alpha = \tan^{-1}(d_o/b_o)$.

The tensile membrane forces on the web are replaced by two forces, each of value F_s, acting above and below the hole

$$2F_s = \sigma_t^y t[2c \sin\theta + (d - b \tan\theta) \cos\theta - \delta] \qquad (2.21)$$

From the vertical equilibrium of forces,

$$2F_s \sin\theta = V_{ult} - (V_{cr})_{red}$$

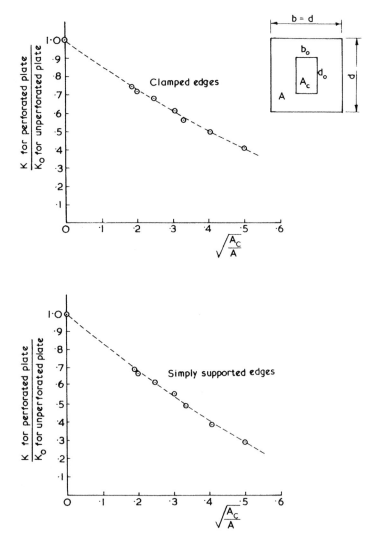

FIG. 2.14. Square plate with central rectangular hole. (From Narayanan and Der Avanessian, 1982a.)

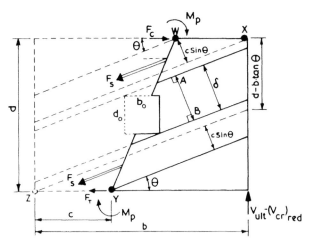

FIG. 2.15.   Forces on the panel.

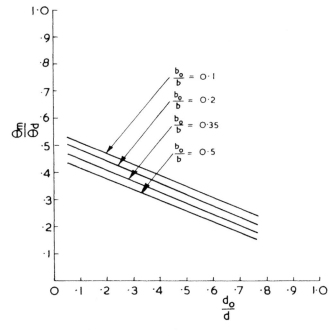

FIG. 2.16.   Variation of optimum angle θ_m with the depth of cutout in a square panel ($b/d = 1.0$; $d/t = 360$).

Substituting for F_s from eqn (2.21), we obtain

$$V_{ult} = 2c\sigma_t^y t \sin^2\theta + \sigma_t^y td(\cot\theta - \cot\theta_d)\sin^2\theta$$
$$- \sigma_t^y t\sqrt{(b_o^2 + d_o^2)}\sin(\alpha + \theta)\sin\theta + (\tau_{cr})_{red}dt \qquad (2.22)$$

Quantities c and c_t^y are evaluated, as before, from eqns (2.2) and (2.11) respectively. Equation (2.22) covers all practical ranges of web holes and is valid for depths of cutouts given by $d_o < [d - (b + b_o)\tan\theta]$.

For a given set of dimensions of plate girder, the only unknown in eqn (2.22) is θ. As this is an equilibrium solution, the maximum value of V_{ult} can be obtained by trial and error, to yield an optimum value θ_m for θ. Figure 2.16 shows the variation of the angle θ_m, with the depth of cutout (d_o/d), for a square panel having $d/t = 360$. Several values of b_o/b are considered and it will be seen that in all cases, a linear relationship between θ_m/θ_d and d_o/d results. Similar parametric studies on several rectangular and square girders having rectangular holes confirm this behaviour.

Openings having depths larger than $(d - (b + b_o)\tan\theta)$ are unlikely to be met in practice and are outside the scope of this paper.

2.6   WEBS CONTAINING REINFORCED CIRCULAR HOLES

If the loss of strength implicit in cutting a web hole is unacceptable, the web will need reinforcing around the hole, so that the opening can be introduced without loss of strength. It would then be necessary to assess the strength of such a girder, with a view to examining its adequacy.

An equilibrium solution for predicting the ultimate capacity of girders with reinforced central circular cutouts is given below and is an extension of the theory suggested for webs with unreinforced circular openings. Figure 2.17 shows a model of the failure mechanism used and is based on patterns of collapse observed in test girders. Once again the solution is based on the equilibrium of the forces in this collapse state.

As before, two tension bands are formed, one above and the other below the cutout. An additional width of the tension field is considered to represent the contribution made by the reinforcement.

Since the boundaries of the cutout are stiffened by the reinforcing ring there will be a uniform tensile membrane stress set up across the web prior to buckling. However, as the load is gradually increased, the ring, together with the web, buckles, and as a result the tensile membrane stresses will start shedding towards the unbuckled regions of the web (see Fig. 2.18(a)).

For simplicity the membrane stresses within the web are replaced by an equivalent width of the tension field, as shown in Fig. 2.18(b). This is, of course, analogous to the concept of effective width of a plate under uniaxial compression.

The additional width b_e of the stress field defines the contribution made by the reinforcement, and its value is related to the cross-section of the ring. Figure 2.19(a) shows the free body diagram of the right half of the plate at

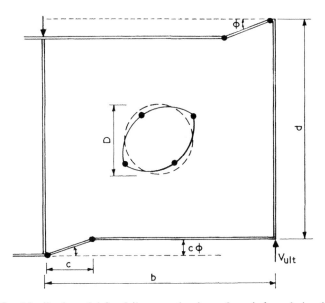

FIG. 2.17.   Idealised model for failure mechanism of a reinforced circular hole.

the instant of collapse. From the equilibrium of forces acting over a quarter of the ring between two plastic hinges (see Fig. 2.19(b)):

$$M_{pr} = \frac{\sigma_t^y b_e^2 t}{16} \tag{2.23}$$

where M_{pr} = plastic moment of resistance of the ring and is given by $M_{pr} = \sigma_{yr}(t_r w_r^2/4)$, b_e = width of the additional tension band, σ_{yr} = yield stress of the reinforcing material, t_r = thickness of reinforcement ring including the web thickness and w_r = width of the reinforcement.

Let us define M_{pr}^* as the moment of resistance of the reinforcing ring

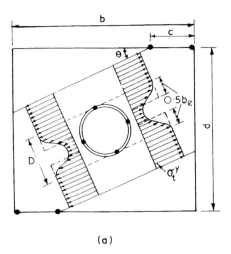

(a)

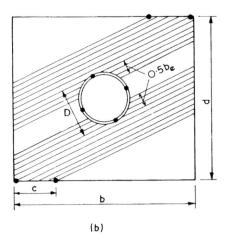

(b)

FIG. 2.18.   (a) Tensile membrane stresses; (b) equivalent tension field.

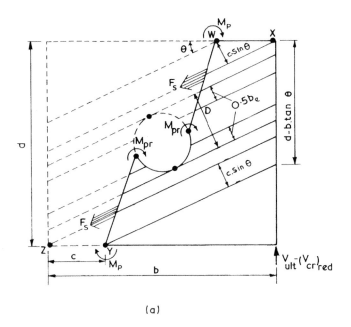

(a)

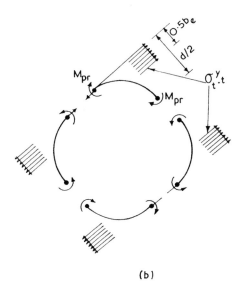

(b)

FIG. 2.19.   (a) Free body diagram; (b) forces on a quarter ring.

which is just capable of producing a uniform stress, σ_t^y, within the web as if there was no cutout (i.e. when $b_e = D$):

$$M_{pr}^* = \frac{\sigma_t^y D^2 t}{16} \tag{2.24}$$

hence

$$\frac{b_e}{D} = \sqrt{\frac{M_{pr}}{M_{pr}^*}} \leq 1 \cdot 0$$

and therefore

$$b_e = \sqrt{\left(\frac{16 M_{pr}}{\sigma_t^y t}\right)} \tag{2.25}$$

The tensile membrane forces are replaced by two forces F_s (see Fig. 2.19(a)). From the geometry of the structure it can be seen that

$$2F_s = \sigma_t^y t \{2c \sin \theta + (d - b \tan \theta) \cos \theta - D\} + \sigma_t^y t b_e$$

Considering the vertical equilibrium of the forces we obtain

$$2F_s \sin \theta = V_{ult} - (V_{cr})_{mod}$$

where $(V_{cr})_{mod} =$ critical shear appropriate to a reinforced web. Substituting for F_s in the above equation, V_{ult} is given by

$$V_{ult} = 2c\sigma_t^y t \sin^2 \theta + \sigma_t^y t d(\cot \theta - \cot \theta_d) \sin^2 \theta$$

$$- \sigma_t^y t D \left(1 - \frac{b_e}{D}\right) \sin \theta + (\tau_{cr})_{mod} dt \tag{2.26}$$

where c and σ_t^y are given by eqns (2.2) and (2.11) respectively. The maximum value of V_{ult} can be obtained by trial and error, as before. It will be noted that when b_e is zero the expression for V_{ult} gives the case for webs with unreinforced circular cutout, and when $b_e = D$ the equation will be the same as for unperforated webs.

Values for $(\tau_{cr})_{mod}$ which are appropriate to the reinforced web should be used in eqn (2.26). This will present some difficulty, as standard solutions for reinforced webs under shear are not available in published literature. Narayanan and Der Avanessian (1982a) used finite element methods of analyses for obtaining elastic critical stresses for various values of t_r/t and w_r/d for two values of D/d. Examples of their results are given in Fig. 2.20.

However, as the purpose of the reinforcement is to restore the stiffness of the web to its full capacity, the elastic critical stresses corresponding to the full web can be used to predict the ultimate loads. This is, in fact, a conservative assumption and safe values of collapse loads are predicted.

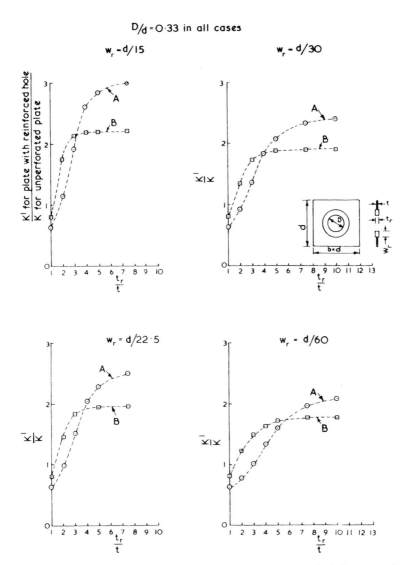

FIG. 2.20.   Variation of the elastic critical stress with the area of reinforcement for buckling modes A and B. (From Narayanan and Der Avanessian, 1982*a*.)

2.7 WEBS CONTAINING REINFORCED RECTANGULAR HOLES

Webs with rectangular holes are usually reinforced by flats or bars welded to both faces of the web and located symmetrically above and below the cutouts; an example of the type of reinforcement considered is seen in Fig. 2.28.

An equilibrium solution for predicting the ultimate capacity of girders containing such reinforced rectangular cutouts is suggested below and is based on the failure mechanisms observed in tests; the method is an extension of the theory suggested for panels with unreinforced rectangular openings. Figure 2.21 shows the hinges formed at the instant of failure in the flanges and on the web reinforcement. For purposes of analysis, two tension bands are considered, one above and the other below the cutout, Fig. 2.22(a). The contribution made by the reinforcement is defined by an additional width of the tension field equal to $2c_r \sin \theta$ (see Fig. 2.22(b)), where c_r is the distance between the hinges on the reinforcement. The position of the hinges will, obviously be related to the cross-sectional area of the reinforcement. Let us consider the free body diagram of the web to

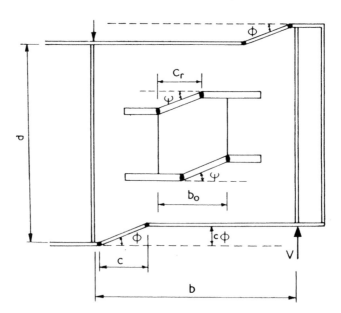

Fig. 2.21.   Idealised model for failure mechanism.

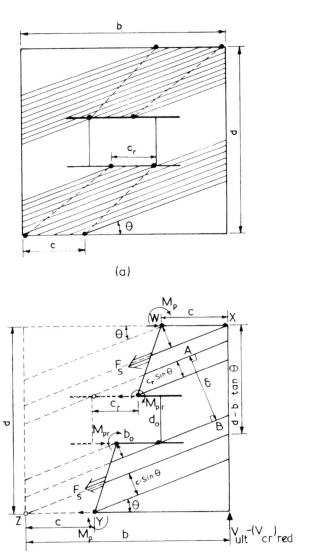

FIG. 2.22.   (a) Formation of hinges; (b) free body diagram.

the right of WY in Fig. 2.22(b). The membrane forces on the web are replaced by two forces (F_s). From the geometry of the structure it can be seen that

$$2F_s = \sigma_t^y t[2c \sin \theta + (d - b \tan \theta) \cos \theta - \delta + 2c_r \sin \theta] \qquad (2.27)$$

where

$$\delta = \sqrt{(b_o^2 + d_o^2)} \sin (\alpha + \theta)$$

Considering the vertical equilibrium of the forces, and noting that there is no shear at the internal hinge positions, V_{ult} is given by:

$$V_{ult} = 2F_s \sin \theta + (V_{cr})_{mod}$$

where $(V_{cr})_{mod}$ = critical shear appropriate to a reinforced web.

$$V_{ult} = 2c\sigma_t^y t \sin^2 \theta + \sigma_t^y td(\cot \theta - \cot \theta_d) \sin^2 \theta$$
$$- \sigma_t^y t\sqrt{(b_o^2 + d_o^2)} \sin (\alpha + \theta) \sin \theta + 2c_r \sigma_t^y t \sin^2 \theta + (\tau_{cr})_{mod} dt \qquad (2.28)$$

where c and σ_t^y are given by eqns (2.2) and (2.11) respectively. If the reinforcement is adequate to restore the full strength of the web, it is appropriate to assume $(\tau_{cr})_{mod}$ to be equal to τ_{cr} for an unperforated web. The method of designing the reinforcement is outlined below.

A certain minimum length of the reinforcement should be welded to the web of the girder, in order to provide adequate end fixity to the reinforcement, Fig. 2.23(a). If the anchorage length l is adequate, then a plastic hinge cannot form at location P (i.e. at the corner of the cutout over the reinforcement); instead, localised yielding will occur over an area of the web associated with the length l, and as a result, the reinforcement will rotate about its *end* and failure would occur as shown in Fig. 2.23(b). The distance between the reinforcement hinges c_{r0}, corresponding to the case when there is an adequate anchorage length l (thus ensuring full-end fixity to the reinforcement) is obtained by considering the equilibrium of the forces acting on the portion of the reinforcement between the plastic hinges:

$$c_{r0} = \frac{2}{\sin \theta} \sqrt{\left(\frac{M_{pr}}{\sigma_t^y t}\right)} \qquad 0 \le c_{r0} \le (b_o + l) \qquad (2.29)$$

where l = anchorage length of the reinforcement and M_{pr} = plastic moment of resistance of the reinforcement, and is given (as before) by:

$$M_{pr} = \sigma_{yr} \frac{t_r w_r^2}{4} \qquad (2.30)$$

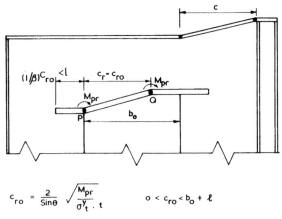

$$c_{ro} = \frac{2}{Sin\theta} \sqrt{\frac{M_{pr}}{\sigma^y_t . t}} \qquad 0 < c_{ro} < b_o + \ell$$

(a) ADEQUATELY ANCHORED REINFORCEMENT

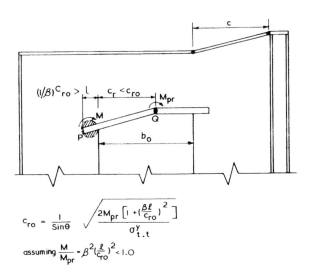

$$c_{ro} = \frac{1}{Sin\theta} \sqrt{\frac{2M_{pr}\left[1 + (\frac{\beta\ell}{c_{ro}})^2\right]}{\sigma^y_t . t}}$$

assuming $\dfrac{M}{M_{pr}} = \beta^2(\frac{\ell}{c_{ro}})^2 < 1.0$

(b) INADEQUATELY ANCHORED REINFORCEMENT

FIG. 2.23.   Reinforcement to the web of the girder.

If the anchorage length l is inadequate, then the reinforcement will rotate at an applied moment smaller in magnitude than M_{pr}. Let M be the moment at which this rotation takes place. Obviously, $M/M_{pr} < 1.0$. The anchorage length l is obviously dependent on the width of the cutout, within certain practical limits. Let

$$b_o = \beta l \tag{2.31}$$

Obviously β should be adequate to ensure that no rotation takes place at location P. Experimental evidence suggests the following empirical relation:

$$\frac{M}{M_{pr}} = \frac{\beta^2 l^2}{c_{r0}^2} \leq 1.0 \tag{2.32}$$

The distance, c_r, between the internal hinge (Q) from the corner of the cutout (P) is found by considering the equilibrium of forces on the reinforcement at the ultimate limit:

$$c_r = \frac{1}{\sin \theta} \sqrt{\left(\frac{2 M_{pr} \left(1 + \left| \frac{\beta l}{c_{r0}} \right|^2 \right)}{\sigma_t^y t} \right)} \tag{2.33}$$

The above equation is true as long as $c_r \leq c_{r0}$; it will be noted that when $\beta l / c_{r0} = 1$, $c_r = c_{r0}$.

From a knowledge of t_r, w_r, l and σ_{yr} in respect of a chosen reinforcement, c_{r0}, M_{pr} and β can be evaluated from eqns (2.29), (2.30) and (2.31) respectively. Hence c_r can be obtained from eqn (2.33) and compared with c_{r0} from eqn (2.29).

A well designed reinforcement will develop hinges such that c_r is at least equal to c_{r0}. Experiments reported by Narayanan et al. (1983) have established that a safe value is $\beta = 4$. In other words, the anchorage length, l, should be at least $0.25b$; in this case eqn (2.29) would give the hinge distance to be substituted in eqn (2.28).

2.8   EXPERIMENTAL INVESTIGATIONS

On account of the lack of published data on the ultimate behaviour of plate girders with web holes, it was necessary to carry out systematically, a number of tests aimed at verifying the theories outlined in the preceding sections. A number of parameters can be expected to influence the ultimate behaviour of webs containing holes; the following are the significant ones:

1. shear force at the cutout;
2. bending moment at the centre line of the cutout;
3. slenderness (d/t) of the web;
4. the hole size relative to the web size;
5. the stiffness of the flange;
6. the presence of any reinforcement around the holes.

The theoretical models and the experiments were designed to ascertain the ultimate capacity of the girders under predominantly shear loading; no studies were, therefore, carried out on webs subjected to high bending moments.

Some 70 ultimate load tests were carried out on girders of various dimensions. The parameters studied included the web slenderness, the hole size, flange stiffness and the influence of the reinforcement. Figures 2.24–2.28 give the design details of the girders tested and Table 2.1 lists the parameters varied. The tests were carried out in a 100-tonne capacity test rig fitted with a Servo-controlled hydraulic system. 'Deflection Control' was employed using an extremely low rate of increment of deflection, so that both the loading and the post-peak unloading behaviour of the test girders could be monitored.

Each test girder consisted of two sections (A and B). The web with the larger opening (say, Panel A), was tested first after stiffening the web which was not under test (Panel B). When the load on web A had reached its ultimate value and the panel had commenced load-shedding, the girder was

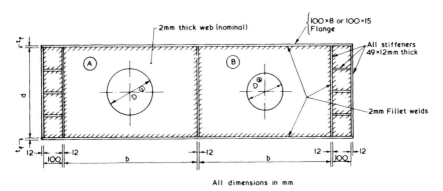

All dimensions in mm

FIG. 2.24.   Panels with centrally placed circular holes.

TABLE 2.1
TEST PROGRAMME

Series	No. of tests	See Fig. No.	Range of d/t values	Range of b/d values	Openings tested	Flange stiffness	Reinforcement
Webs having central circular holes	20	2·24	250–360	1·0, 1·5	$D/d = 0.1$–0.9	Varied	No
Webs having eccentric circular holes	8	2·25	250–360	1·0, 1·5	$D/d = 0.2$–0.67	A single flange size used	No
Webs having central rectangular openings	12	2·26	250–360	1·0, 1·5	$d_o/d = 0.2$–0.75 $b_o/b = 0.15$–0.5	A single flange size used	No
Webs having central circular holes with reinforcement	22	2·27	250–360	1·0, 1·5	$D/d = 0.3$–0.5	A single flange size used	Various sizes
Webs having central rectangular holes with reinforcement	8	2·28	360	1·0	$d_o/d = 0.33$ $b_o/b = 0.22$–0.5	A single flange size used	Various sizes

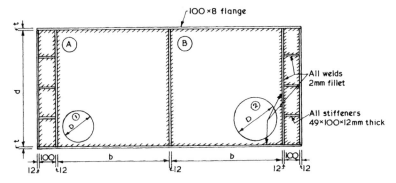

FIG. 2.25. Panels with eccentrically placed holes.

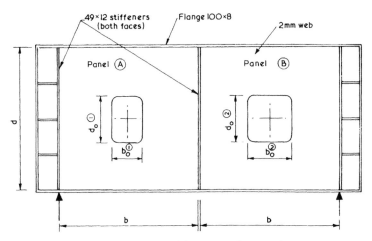

FIG. 2.26. Panels with rectangular openings.

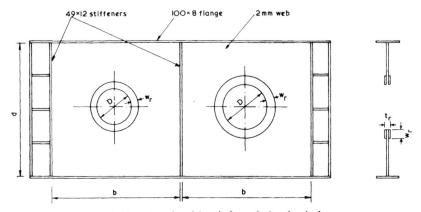

FIG. 2.27. Panels with reinforced circular holes.

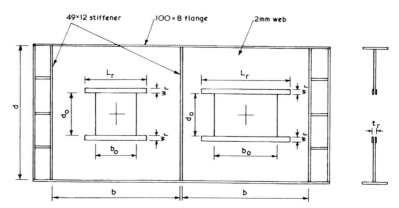

FIG. 2.28. Panels with reinforced rectangular holes.

unloaded, and then Panel B was tested in a similar manner. After the tests, the profiles of the webs and the flanges were obtained using a ripple scanner (Narayanan and Rockey, 1981) so that the locations of the plastic hinges formed at the ultimate stage could be obtained.

The theoretical predictions of the ultimate loads in the girders were obtained by using the measured yield stresses of the web and flange material. These theoretical predictions were invariably found to be safe and very close to the experimentally observed values; a summary of these comparisons is given in Table 2.2.

TABLE 2.2
CALIBRATION OF FORMULAE

Series	No. of tests	Mean value of $\left(\dfrac{Predicted\ load}{Observed\ load}\right)$	Standard deviation
Webs having central circular holes	20	0·84	0·067
Webs having eccentric circular holes	8	0·91	0·055
Webs having central rectangular holes	12	0·85	0·080
Webs having central circular holes with reinforcement	22	0·83	0·050
Webs having central rectangular holes with reinforcement	8	0·88	0·038

2.9 CONCLUSIONS

Theoretical models for predicting the ultimate capacity of plate girders containing circular and rectangular web holes are presented. The ultimate strength of such a plate girder is determined as the sum of the elastic critical load, the post-critical membrane tension in the web, the load taken by the flanges and the capacity of the reinforcement provided, if any. Approximate expressions for estimating the elastic critical loads have been suggested. Satisfactory methods for calculating the collapse loads are proposed and are found to be adequate to predict the strengths within acceptable engineering accuracy.

REFERENCES

EUROPEAN CONVENTION FOR CONSTRUCTIONAL STEELWORK (1976) 2nd International Colloquium on Stability, Introductory Report (Aug.), pp. 178–208.

HOGLUND, T. (1971) Strength of thin plate girders with circular or rectangular web holes without web stiffeners. *Proceedings of the London Colloquium of the International Association of Bridge and Structural Engineering.*

NARAYANAN, R. (1980) Ultimate capacity of plate girders containing cutouts. University College, Cardiff, Report.

NARAYANAN, R. and ADORISIO, D. (1983) Model studies on plate girders. *Journal of Strain Analysis*, **18**(2), 111–17.

NARAYANAN, R. and DER AVANESSIAN, N. G. V. (1981a) A theoretical method for the prediction of ultimate capacity of webs with circular cutouts. University College, Cardiff, Report.

NARAYANAN, R. and DER AVANESSIAN, N. G. V. (1981b) Theoretical methods for the assessment of ultimate capacity of plate girders containing central rectangular cutouts and eccentric circular cutouts. University College, Cardiff, Report.

NARAYANAN, R. and DER AVANESSIAN, N. G. V. (1982a) Elastic buckling of perforated plates under shear. University College, Cardiff, Report.

NARAYANAN, R. and DER AVANESSIAN, N. G. V. (1982b) Ultimate capacity of plate girders containing holes in webs. University College, Cardiff, Report.

NARAYANAN, R. and ROCKEY, K. C. (1981) Ultimate capacity of plate girders with webs containing circular cutouts. *Proceedings of the Institution of Civil Engineers*, London, Part 2, **72**, 845–62.

NARAYANAN, R., DER AVANESSIAN, N. G. V. and GHANNAM, M. M. M. (1983) Small scale model tests on perforated webs. Accepted for publication in *Structural Engineer*.

PORTER, D. M., ROCKEY, K. C. and EVANS, H. R. (1975) The collapse behaviour of plate girders loaded in shear. *Structural Engineer*, **53**(8), 315–25.

REDWOOD, R. G. (1972) Tables for plastic design of beams with rectangular holes. *Engineering Journal, American Institute of Steel Construction*, 9(1), 2–19.

REDWOOD, R. G. (1983) Design of I beams with web perforations. *Beams and Beam Columns—Stability and Strength* (Ed. by R. Narayanan), Chapter 3, Applied Science Publishers, London, pp. 95–133.

REDWOOD, R. G. and SHRIVASTAVA, S. C. (1980) Design recommendations for steel beams with web holes. *Canadian Journal of Civil Engineering*, 7(4), 642–50.

ROCKEY, K. C., ANDERSON, R. G. and CHEUNG, Y. K. (1967) The behaviour of square shear webs having a circular hole. *Proceedings of the Swansea Symposium on Thin Walled Structures*, Crosby, Lockwood and Sons, London, pp. 148–69.

SHANMUGAM, N. E. and NARAYANAN, R. (1982) Elastic buckling of perforated plates for various loading and edge conditions. *International Conference on Finite Element Methods*, Shanghai, Paper No. 103.

TIMOSHENKO, S. P. and GERE, J. M. (1961) *Theory of Elastic Stability*, McGraw-Hill–Kogakusha Ltd, Tokyo.

Chapter 3

PATCH LOADING ON PLATE GIRDERS

T. M. ROBERTS

*Department of Civil and Structural Engineering,
University College, Cardiff, UK*

SUMMARY

*During the past 20 years attention has been directed towards investigating
the ultimate load-carrying capacity of slender plate girders subjected to
localised edge loading or patch loading. Early tests revealed that the collapse
load is dependent primarily upon the square of the web thickness, all other
parameters being of only minor significance, and a large number of tests
carried out since that time have confirmed these conclusions. In this chapter,
various aspects of the problem are discussed including the test programmes,
the analysis of elastic critical loads of idealised web panels and empirical and
theoretical predictions of the collapse loads of both stocky and slender plate
girders. A number of simple closed-form solutions are now available which
enable designers to predict collapse loads with reasonable accuracy.*

NOTATION

b	Width of web panel
b_f	Width of flange
c	Length of patch load
d	Depth of web panel
t	Thickness
t_f	Thickness of flange
t_w	Thickness of web

| u, v, w | Displacements in x, y and z directions |
| x, y, z | Coordinate axes |

D	Flexural rigidity of plate $Et^3/12(1-v^2)$
E	Young's modulus
H	Function of girder dimensions and material properties
I_f	Second moment of area of flange
K	Buckling coefficient
M_f	Plastic moment of flange
M_w	Plastic moment of web
N_x, N_y, N_{xy}	Membrane direct and shearing forces
P	Load
P_{cr}	Elastic critical load
P_{ex}	Experimental collapse load
P_{pr}	Predicted collapse load
P_u	Ultimate or collapse load
P_{ub}	Ultimate load—web handling
P_{uy}	Ultimate load—web yielding

α	Dimension defining collapse mechanism
β	Dimension defining collapse mechanism
γ	b/d
δ	Small variation
ε_y	Membrane strain in y-direction
θ	Angle defining deformation of web
λ	Function of girder dimensions and material properties
v	Poisson's ratio
σ	Stress
σ_b	Bending stress
σ_f	Yield stress of flange
σ_w	Yield stress of web
σ_y	Yield stress

3.1 INTRODUCTION

Patch loading or localised edge loading of slender plate girders is a problem
frequently encountered in practice. Examples of this type of loading are
wheel loads on gantry girders, loads from purlins onto the main frame
member of buildings and roller loads during the launching of plate and box

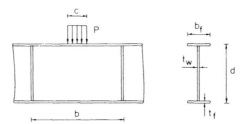

FIG. 3.1.   Patch loading and girder dimensions.

girders. In bridge construction, where plate girders support relatively thin concrete or steel deck slabs, the plate girders may be subjected to localised wheel loading distributed through the deck slabs.

The type of loading under consideration is illustrated in Fig. 3.1. In some situations the loaded length c may be small enough for the load to be assumed concentrated at a point while in other situations the loading may extend the entire distance between vertical web stiffeners. This latter type of loading will be referred to as distributed edge loading.

The localised stress distribution due to the patch loading will be, in general, combined with global bending and shear stresses. Whether it is the local or the global stress distribution which is predominant will depend on the overall structural form and loading.

Although the problem of patch loading may seem to be of only minor significance in the overall design, it is however, important to ensure that girders are not overstressed locally since localised failure may precipitate overall structural failure. The simplest design procedure for localised loading is to limit the magnitude of the applied load to a proportion of the contact area times the yield stress of the material. This may prove to be a satisfactory design procedure for girders with relatively thick stocky webs but in general does not show any correlation with the actual failure loads of girders obtained from model tests.

During the past 20 years a large number of model tests have been performed by several research workers to provide a better understanding of the mode of failure, and influence of the geometric and material parameters on the failure load of slender plate girders subjected to patch loading. The mode of failure observed in nearly all of the tests is as shown in Fig. 3.2. Failure occurs due to the formation of plastic hinges in the flange accompanied by yield lines in the web. This type of failure, which is very localised, has become known as web crippling.

Theoretical investigations have concentrated on two main aspects of the

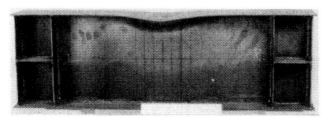

FIG. 3.2.   Typical mode of failure of plate girder subjected to patch loading.

problem. Rigorous analytical and numerical solutions for the elastic critical loads of web panels, with assumed idealised boundary conditions and combined in-plane loading, have been obtained. These solutions show little or no correlation with experimental failure loads. This is due to the post-buckled reserve of strength possessed by restrained thin panels and the interaction between the web and the flanges. Elastic critical loads still have significance however in design for service conditions.

An alternative approach has been to base the prediction of failure loads on simple empirical and semi-empirical formulae. Many such formulae, differing only slightly in form, have been proposed and failure loads (for model tests) can be predicted to within $\pm 20\%$. Considering the simplicity of many of these formulae and the complex material and geometrically nonlinear nature of the problem, these predictions are satisfactory for practical purposes.

Recent theoretical investigations have been based on assumed collapse mechanisms, the appropriate mechanisms being deduced from experimental observations. Although such an approach appears in accordance with the upper-bound theorem of plastic collapse, it should be appreciated that plastic-bound theorems are not valid for geometrically nonlinear problems and a strict adherence to these theorems is not theoretically justifiable. Solutions based on these assumed mechanisms of collapse have been reduced to simple closed forms similar to many of the semi-empirical formulae.

3.2   EXPERIMENTAL RESULTS

In the early 1960s, Granholm carried out a series of tests on slender plate girders to investigate the behaviour of webs when subjected to bending, shear and localised edge loading. The original reports, which were written in Swedish, have been summarised by Bergfelt (Granholm, 1976). Based on these test results, Granholm proposed the simple formula

$$P_u = 0.85 t_w^2 \qquad (3.1)$$

for predicting the ultimate load for web crippling, P_u, in terms of the web thickness, t_w. In this formula, which is dimensionally dependent, P_u is in tonnes and t_w is in millimetres. It is surprising, when considering the complex nature of the problem, that such a simple formula should give a satisfactory prediction of the ultimate load. However, experimental and theoretical investigations carried out over the past 20 years have confirmed the dependance of P_u on the square of the web thickness, and that other parameters are of only minor significance. Equation (3.1) has been for many years incorporated in the Swedish Standard for the design of plate girders.

Following on from these early tests, several researchers have carried out extensive experimental investigations of the influence of other parameters such as the web depth, d, the spacing between vertical web stiffeners, b, the flange width, b_f, and the flange thickness, t_f (Bergfelt and Hovik, 1968, 1970; Bergfelt, 1971, 1979; Bergfelt and Lindgren, 1974; Škaloud and Novak, 1975; Drdacky and Novotny, 1977; Roberts and Rockey, 1979; Roberts, 1981a). The majority of the available test data has been summarised by Roberts (1981a) and compared with current predictions. Before proceeding with a theoretical discussion of the problem it is of interest to examine the general behaviour observed and conclusions reached by all investigators.

Tests have been performed on short and medium span girders, with and without vertical web stiffeners, and with web thicknesses between 1 and 4 mm. At present, very little test data is available for girders with web thicknesses greater than 4 mm. This is somewhat surprising since web thickness appears to be the most significant parameter. The collapse or crippling load varies approximately as the square of the web thickness. It is influenced to a much lesser extent by the length of the patch load, the flange thickness, the spacing of vertical web stiffeners and the material yield stress but is almost independent of the depth of the web and the width of the flange.

Details of a number of short-span girders tested by Roberts (1980, 1981b) are given in Table 3.1. The dimensions and loading are defined in Fig. 3.1 and σ_w and σ_f are the yield stresses of the web and flange respectively; P_{ex} is the maximum value of the applied load reached during a test. These results have been chosen for discussion since they are representative of a wide

T. M. ROBERTS

TABLE 3.1

Girder	b (mm)	d (mm)	t_w (mm)	b_f (mm)	t_f (mm)	c (mm)	σ_w (N/mm²)	σ_f (N/mm²)	P_{ex} (kN)
B2-3	600	500	2·12	150	3·05	50	224	221	34·08
B2-7	600	500	2·12	150	6·75	50	224	279	37·92
B2-12	600	500	2·12	150	11·75	50	224	305	44·16
B2-20	600	500	2·12	150	20·06	50	224	305	84·48
E10-10-1/1	500	500	9·95	150	10·05	0	222	240	716·0
E10-10-2/1	500	500	9·95	150	10·05	100	247	250	787·0

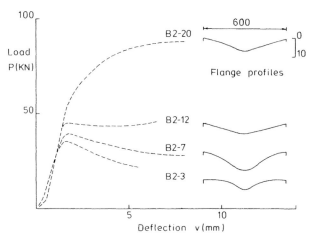

FIG. 3.3.   Load deflection curves and flange profiles for the B-series girders.

range of web thickness. The results obtained for each test were the curves of load P versus corresponding deflection v, profiles of the out-of-plane deflection of the web for various values of the applied load and horizontal flange profiles at the end of the test. Load–deflection curves and residual flange profiles are shown in Fig. 3.3 for the B-series girders and in Fig. 3.4 for the E-series girders. The web profiles indicated that out-of-plane deflection of the web became pronounced, only when the load approached its maximum value.

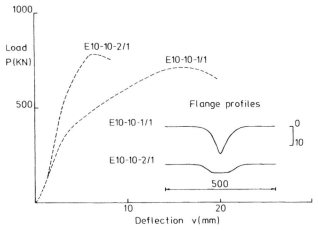

FIG. 3.4.   Load deflection curves and flange profiles for the E-series girders.

For the B-series girders, the nature of the load–deflection curves depends upon the flange thickness. All curves are initially linear. For relatively thin flanges failure occurs quite suddenly and is characterised by a rapid growth in the out-of-plane deflections of the web and a significant fall in the load-carrying capacity. For very thick flanges, the load–deflection curves show distinct nonlinearity, long before the maximum load is reached. The load then increases gradually to a maximum and remains approximately constant for increasing deflection. These load–deflection curves appear to be influenced by the position at which the plastic hinges in the flange are formed. For thin flanges the hinges form close to the load and failure occurs quite suddenly. For increasing flange thickness the point at which the outer plastic hinges form moves outwards until eventually the plastic hinges are constrained to form adjacent to the vertical web stiffeners. This restriction of the spread of the plastic hinges in the flange results in the characteristic load–deflection curve shown for girder B2-20.

The load–deflection curves for the E-series girders, which had much thicker webs, differ from those for the B-series girders. The curves are approximately linear up to 40–50 % of the maximum load. The curves then exhibit distinct nonlinearity after which the load increases gradually to a maximum. Strain gauge readings showed the nonlinearity in the load–deflection curves to be due to membrane yielding of the web beneath the load, long before the maximum load was reached. Even so, ultimate failure was due to web crippling, characterised by a rapid growth in the out-of-plane deflection of the web. The horizontal flange profiles indicate that the plastic hinges in the flange formed quite close to the load.

3.3 ELASTIC CRITICAL LOADS OF WEBS

It has already been mentioned that the elastic critical loads of webs show little or no correlation with failure loads determined from model tests due to the post-buckled reserve of strength possessed by restrained thin panels and the interaction between the web and the flanges. Elastic critical loads still have significance however in design for service conditions. The particular problem under consideration also highlights the errors which may be introduced in the calculation of elastic critical loads by the assumption of inextensional buckling.

Sommerfield (1906), Timoshenko (1910) and Timoshenko and Gere (1961) provided the first approximate solutions to the problem of the buckling of a simply-supported plate compressed by two equal and

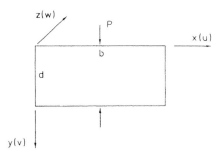

FIG. 3.5. Simply-supported plate.

opposite forces as shown in Fig. 3.5. If displacements in the x, y and z directions are denoted by u, v and w respectively, the energy equation for plate buckling is (Timoshenko, 1961):

$$\frac{D}{2} \int \int \left\{ \left(\frac{\partial^2 \delta w}{\partial x^2} + \frac{\partial^2 \delta w}{\partial y^2} \right)^2 - 2(1-v) \left[\frac{\partial^2 \delta w}{\partial x^2} \frac{\partial^2 \delta w}{\partial y^2} - \left(\frac{\partial^2 \delta w}{\partial x \partial y} \right)^2 \right] \right\} dx\, dy$$

$$+ \frac{1}{2} \int \int \left\{ N_x \left(\frac{\partial \delta w}{\partial x} \right)^2 + N_y \left(\frac{\partial \delta w}{\partial y} \right)^2 + 2 N_{xy} \frac{\partial \delta w}{\partial x} \frac{\partial \delta w}{\partial y} \right\} dx\, dy = 0 \quad (3.2)$$

In eqn (3.2), D is the flexural rigidity of the plate, v is Poisson's ratio and δw is a small variation in the displacement w which occurs due to buckling. N_x, N_y and N_{xy} (tensile +ve) are the membrane direct and shear forces per unit length prior to buckling which for the problem under consideration, and in general, vary throughout the plate.

To obtain a solution Timoshenko replaced the second integral in eqn (3.2), which represents the change in membrane energy associated with the buckling displacement δw, by

$$-\frac{P}{2} \int_0^d \left(\frac{\partial \delta w}{\partial y} \right)_{x=b/2}^2 dy \quad (3.3)$$

This approximation can be viewed in two ways. It is either equivalent to assuming that the membrane forces are concentrated along the line of action of the external forces or it is equivalent to the assumption of inextensional buckling. To understand this latter aspect it is necessary to consider the nonlinear expression for the variation in the membrane strain in the y-direction due to buckling, $\delta \varepsilon_y$ which is

$$\delta \varepsilon_y = \frac{\partial \delta v}{\partial y} + \frac{1}{2} \left(\frac{\partial \delta w}{\partial y} \right)^2 \quad (3.4)$$

If $\delta\varepsilon_y$ is assumed equal to zero throughout the plate, then

$$P \int_0^d (\delta\varepsilon_y)_{x=b/2} \, dy = P \int_0^d \left\{ \frac{\partial \delta v}{\partial y} + \frac{1}{2} \left(\frac{\partial \delta w}{\partial y} \right)^2 \right\} \, dy = 0 \qquad (3.5)$$

Hence from eqn (3.5)

$$P\{(\delta v)_{\substack{y=0 \\ x=b/2}} - (\delta v)_{\substack{y=d \\ x=b/2}}\} = \frac{P}{2} \int \left(\frac{\partial \delta w}{\partial y} \right)^2 \, dy \qquad (3.6)$$

The left-hand side of eqn (3.6) represents the work done by the forces P during the buckling displacements δv. If it is now assumed that as buckling occurs there is no change in the membrane strains the energy equation for plate buckling can be obtained by equating the work done by the external forces to the change in the bending energy of the plate. Hence

$$\frac{D}{2} \int \int \left\{ \left(\frac{\partial^2 \delta w}{\partial x^2} + \frac{\partial^2 \delta w}{\partial y^2} \right)^2 - 2(1-v) \left[\frac{\partial^2 \delta w}{\partial x^2} \frac{\partial^2 \delta w}{\partial y^2} - \left(\frac{\partial^2 \delta w}{\partial x \, \partial y} \right)^2 \right] \right\} dx \, dy$$

$$- \frac{P}{2} \int_0^d \left(\frac{\partial \delta w}{\partial y} \right)^2_{x=b/2} \, dy = 0 \qquad (3.7)$$

Timoshenko obtained a solution of eqn (3.7) by assuming δw in the form of an infinite series, and showed that the solution for the critical load P_{cr} could be expressed as

$$P_{cr} = K \frac{\pi^2 D}{d} \qquad (3.8)$$

where K is a nondimensional buckling parameter given by

$$K = \frac{1}{2\gamma^3 \sum_{n=1,3,5} (\gamma^2 + n^2)^{-2}} ; \qquad \gamma = b/d \qquad (3.9)$$

A more rigorous solution of this problem was presented by Leggett (1937) who used a stress function in the form of an infinite series to represent the membrane forces in the plate prior to buckling. Leggett's solution has been confirmed using the finite element method and the results indicate that Timoshenko's approximate solution is approximately 25 % in error for plates having b/d equal to unity. The error reduces however to approximately 12 % for long plates. A plot of the K values obtained from eqn (3.9) and using the finite element method is shown in Fig. 3.6.

Recently, Khan and Walker (1972) obtained solutions for the problem illustrated in Fig. 3.7, where the applied loads are distributed over a finite

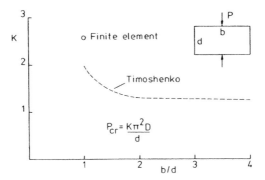

FIG. 3.6.   Variation of buckling coefficient K with b/d ratio.

length c. Khan and Walker used a modified form of the energy equation for plate buckling, suggested by Alfutov and Balabukh (1967, 1968), in which it is not necessary to incorporate an exact representation of the membrane stresses prior to buckling. The assumed approximate stress distribution should satisfy the equations of equilibrium throughout the plate and the boundary conditions. The formulation is made rigorous by the inclusion of a stress function which must satisfy clearly-defined conditions throughout the plate and on the boundaries. The buckling coefficient K obtained for a square plate and varying values of c/d is shown in Fig. 3.7.

Girkmann (1936) was the first investigator to study the problem of the buckling of a rectangular plate subjected to a single-edge load as shown in

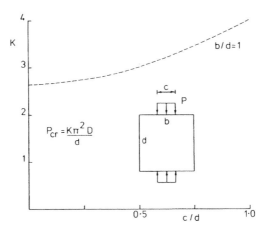

FIG. 3.7.   Variation of buckling coefficient K with c/d ratio for a square plate.

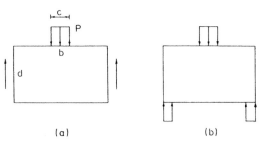

FIG. 3.8. Loading and support conditions for buckling of a rectangular plate subjected to a single-edge load (see Section 3.3 for explanation of (a) and (b)).

Fig. 3.8. Zetlin (1955) provided a more detailed study of the problem using energy methods and presented his data in graphical form. Zetlin assumed the plate to be simply-supported along all four edges and that the applied load was supported by shear stresses, distributed parabolically over the two ends as shown in Fig. 3.8(a). White and Cottingham (1962) examined the buckling of a web plate when loaded and supported as shown in Fig. 3.8(b) using a finite difference solution. Results were presented for simply-supported and clamped boundaries. Rockey and Bagchi (1970) solved similar problems using the finite element method.

In a series of papers (Khan and Walker, 1972; Khan and Johns, 1975; Khan et al., 1977) the modified energy equation proposed by Alfutov and Balabukh (1967, 1968), was used to study the buckling of plates subjected to a variety of loading conditions. Localised edge loading with combined

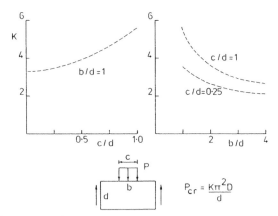

FIG. 3.9. The K values for a simply-supported plate (loaded and supported as in Fig. 3.8(a)) for various c/d and b/d ratios.

bending and shear stresses was considered and results were presented in the form of interaction curves for the buckling coefficient K. The K values for a simply-supported plate, loaded and supported as in Fig. 3.8(a) are shown in Fig. 3.9 for various c/d and b/d ratios.

If the longitudinal edges of the plate are assumed to be clamped instead of simply-supported, the buckling coefficient increases significantly. This assumption is not however justified when using these idealised results for isolated plates to predict the buckling loads for the webs of welded plate girders. This is due to the imperfections and residual stresses introduced along the web flange junction by the welding and the uncertain restraint offered by the flanges.

3.4 PREDICTION OF COLLAPSE LOADS

During the last decade, many simple formulae have been proposed for predicting the collapse load of plate girders subjected to patch loading. These formulae emphasise the significance of the web thickness, as in eqn (3.1) proposed by Granholm (1976), and the relative insignificance of all other parameters.

Bergfelt (1971) investigated a simple analytical model based on the analogy of a beam on an elastic foundation, the flange being the beam and the web the elastic foundation. The analysis was centred on the determination of an elastic length: this being the distance between points of maximum moment in the flange. The difficulty with this approach was how to specify the equivalent spring stiffness of the web, which is required to calculate the elastic length. Bergfelt also recognised the unsatisfactory nature of eqn (3.1) in that it is dimensionally dependent, and proposed an alternative empirical equation

$$P_u = 4\cdot5 \times 10^{-2} E t_w^2 \qquad (3.10)$$

in which E is Young's modulus. Based on a limited number of test results he also proposed that the influence of coexistent bending stress, σ_b, on the crippling load could be allowed for by reducing the value of P_u given by eqn (3.10) by a factor

$$\left\{1 - \left(\frac{\sigma_b}{\sigma_w}\right)^2\right\}^{1/8} \qquad (3.11)$$

Bergfelt and Lindgren (1974) extended the theoretical discussion based on the analogy of a beam on an elastic foundation to include various assumptions concerning the equivalent spring stiffness of the web. The resulting equations were not generally completely defined but indicated a suitable parametric form. A number of equations were proposed, only one of which is reproduced herein.

$$P_u = 0.68t_w^2(E\sigma_w)^{0.5}(t_f/t_w)^{0.6}f(c) \tag{3.12}$$

In the original equation, $f(c)$ was not defined but was included to allow for the influence of the loaded length c.

Dubas and Gehri (1975) proposed the empirical formula

$$P_u = 0.75t_w^2 \left\{ E\sigma_w \frac{t_f}{t_w} \right\}^{0.5} \tag{3.13}$$

and Škaloud and Drdacky (1975) proposed the following empirical formula which is now included in the Czechoslovak design rules for plate girders

$$P_u = 0.55t_w^2 \left\{ E\sigma_w \frac{t_f}{t_w} \right\}^{0.5} (0.9 + 1.5c/d) \left\{ 1 - \left(\frac{\sigma_b}{\sigma_w} \right)^2 \right\}^{0.5} \tag{3.14}$$

Roberts and Rockey (1977, 1979) and Roberts (1981a) obtained solutions for the collapse load based on assumed mechanisms of collapse. These mechanism solutions are discussed in detail in the next section and only the final simplified equations are given here. It was proposed that the collapse load be taken as the lesser of

$$P_u = 2\{4M_f\sigma_w t_w\}^{0.5} + \sigma_w t_w c \tag{3.15}$$

and

$$P_u = 0.5t_w^2 \left\{ E\sigma_w \frac{t_f}{t_w} \right\}^{0.5} \left\{ 1 + \frac{3c}{d} \left(\frac{t_w}{t_f} \right)^{1.5} \right\} \tag{3.16}$$

In eqn (3.15), M_f is the plastic moment of resistance of the flange. To allow for coexistent bending stress σ_b, the value of P_u given by eqns (3.15) and (3.16) should be reduced by a factor

$$\left\{ 1 - \left(\frac{\sigma_b}{\sigma_w} \right)^2 \right\}^{0.5} \tag{3.17}$$

An equation similar to eqn (3.15) was proposed by Bergfelt (1979) in a paper in which his early work was formalised and named the 'three hinge flange theory', although it is not obvious that such an equation was implied by his earlier work (Bergfelt, 1971).

All these equations, except eqn (3.15), give satisfactory agreement with failure loads obtained from model tests on girders with thin webs, i.e. less than 5 mm. For such girders, eqn (3.15) overestimates the collapse load. It will be shown later however that eqn (3.15) has significance for girders with thicker webs, and therefore for girders of practical dimensions, since it provides a lower-bound solution for the load at which membrane yielding of the web becomes pronounced.

3.5 MECHANISM SOLUTIONS

In this section the mechanism solutions for predicting the collapse or crippling load of plate girders subjected to patch loading, presented by Roberts and Rockey (1977, 1979) and Roberts (1981a), are discussed in detail.

3.5.1 Mechanism Solution—Web Bending

Experimental evidence has shown conclusively that crippling occurs in the manner indicated in Fig. 3.2. Plastic hinges form in the loaded flange accompanied by yield lines in the web. Although the plastic hinges and yield lines are not always as clearly defined as those in Fig. 3.2 the available test data supports the choice of the collapse mechanism shown in Fig. 3.10. Dimensions α and β define the position of the assumed yield lines in the web and plastic hinges in the flange and the angle θ defines the deformation of the web just prior to collapse. Stretching of the web is neglected since for thin panels, energy associated with stretching is much greater than energy associated with bending. M_w, σ_w, M_f and σ_f are the plastic moments and yield stresses of the web and flange respectively (plastic moment per unit length of web).

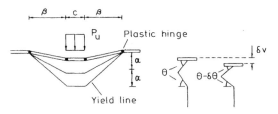

FIG. 3.10. Collapse mechanism—web bending.

If the applied load moves vertically through a small distance δv, the rotation, $\delta \theta$, of the plastic hinges in the flange is $\delta v/\beta$ and of the yield lines in the web is $\delta v/2\alpha \cos \theta$ (twice this value along the central yield line). Equating external and internal work gives

$$P_u = \frac{4M_f}{\beta} + \frac{4\beta M_w}{\alpha \cos \theta} + \frac{2cM_w}{\alpha \cos \theta} \qquad (3.18)$$

and minimising P_u with respect to β only gives

$$\beta^2 = \frac{M_f \alpha \cos \theta}{M_w} \qquad (3.19)$$

An estimate of the deflection of the flange just prior to collapse can be made using elastic theory. If it is assumed that the moment in the flange varies linearly from $+M_f$ at the outer hinge to $-M_f$ at the inner hinge, the deflection of the flange is given by $M_f\beta^2/6EI_f$ where E is Young's modulus and $I_f = b_f t_f^3/12$ is the second moment of area of the flange about its centroidal axis. This deflection must be compatible with the deformation of the web given by $2\alpha(1 - \sin \theta)$. Hence

$$\frac{M_f\beta^2}{6EI_f} = 2\alpha(1 - \sin \theta) \qquad (3.20)$$

and substituting for β^2 from eqn (3.19) gives

$$\frac{\cos \theta}{1 - \sin \theta} = \frac{4E\sigma_w t_w^2}{\sigma_f^2 b_f t_f} \qquad (3.21)$$

When the spread of the plastic hinges in the flange is restricted by closely-spaced vertical web stiffeners the value of β determined from eqn (3.19) cannot exceed $0{\cdot}5(b - c)$. Therefore, β is assigned the value $0{\cdot}5(b - c)$ and from eqn (3.20)

$$\sin \theta = 1 - \frac{M_f\beta^2}{12EI_f\alpha} \qquad (3.22)$$

If α is now chosen empirically, eqns (3.18)–(3.22) lead to a value for the collapse load.

The mechanism solution can be reduced to a simple closed form as follows. Equation (3.21) can be expressed as

$$\frac{\cos \theta}{1 - \sin \theta} = H \qquad (3.23)$$

in which H is a known function of the girder dimensions and material properties. Solving eqn (3.23) for $\cos \theta$ in terms of H gives

$$\cos \theta = \frac{2H}{1 + H^2} \simeq \frac{2}{H} \qquad (3.24)$$

since H is large compared with unity. Substituting eqns (3.24) and (3.19) into eqn (3.18) then gives

$$P_{u} = 2\sqrt{2}t_w^2 \left\{ \frac{E\sigma_w^2 t_f}{\sigma_f \alpha} \right\}^{0.5} \left\{ 1 + \lambda c \left(\frac{t_w}{t_f} \right)^{1.5} \right\} \qquad (3.25)$$

in which λ is a complex function of the girder dimensions and material properties.

Equation (3.25) indicates a slight anomaly in the mechanism solution since if the yield stress of the flange is increased, P_u decreases. This anomaly can be rectified by assuming σ_f is approximately equal to σ_w and hence cancelling σ_f from eqn (3.25).

In the original mechanism solution (Roberts and Rockey, 1977, 1979) α was assumed to be a function of both the depth and thickness of the web. This is not however in accordance with experimental evidence, in particular that presented by Bergfelt (1979) and Roberts (1980), which indicates that for slender plate girders, P_u is independent of the depth. Therefore, in a later version of the mechanism solution (Roberts, 1981a), α was taken simply as

$$\alpha = 25t_w \qquad (3.26)$$

Equation (3.25) now reduces to

$$P_u = 0{\cdot}56t_w^2 \left\{ E\sigma_w \frac{t_f}{t_w} \right\}^{0.5} \left\{ 1 + \lambda c \left(\frac{t_w}{t_f} \right)^{1.5} \right\} \qquad (3.27)$$

To maintain the simplicity of eqn (3.27) the complex function λ is replaced by $3/d$ and so that eqn (3.27) provides a lower-bound solution for all available test data, the constant is reduced from 0·56 to 0·5. Hence

$$P_u = 0{\cdot}5t_w^2 \left\{ E\sigma_w \frac{t_f}{t_w} \right\}^{0.5} \left\{ 1 + \frac{3c}{d} \left(\frac{t_w}{t_f} \right)^{1.5} \right\} \qquad (3.28)$$

Equation (3.28) recognises that the influence of the loaded length c diminishes rapidly with increasing flange thickness. When c/d becomes large, it is unrealistic to assume that the flange remains straight between the two inner plastic hinges and the geometric considerations are suspect. It is recommended therefore that the value of c/d used in eqn (3,28) be limited to 0·2.

3.5.2 Mechanism Solution—Web Yielding

As the web thickness is increased the ratio of the out-of-plane bending stiffness to the compressive membrane stiffness also increases. It might be expected therefore that failure of stocky girders would be initiated by direct yielding of the web and this situation can be analysed by considering the alternative failure mechanism shown in Fig. 3.11. It is assumed that plastic

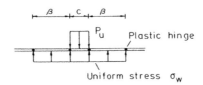

FIG. 3.11.   Collapse mechanism—web yielding.

hinges form in the flange and that the length of web between the outer plastic hinges yields in compression. Equating external and internal work as the load moves vertically through a small distance δv gives

$$P_u = \frac{4M_f}{\beta} + \sigma_w t_w (\beta + c) \qquad (3.29)$$

Energy dissipation around the boundary of the assumed plastic region is neglected. Minimising P_u with respect to β gives

$$\beta^2 = \frac{4M_f}{\sigma_w t_w} \qquad (3.30)$$

and substituting for β in eqn (3.24) gives

$$P_u = 2(4M_f \sigma_w t_w)^{0.5} + \sigma_w t_w c \qquad (3.31)$$

3.5.3 Influence of Coexistent Bending Stress

In practice, web panels of plate girders are often subjected to edge loading and combined bending. The influence of the coexistent bending stress σ_b on the collapse load can be assessed from eqn (3.28). The plastic moment of resistance of a rectangular section is reduced in the presence of coexistent axial stress σ by a factor $1 - (\sigma/\sigma_y)^2$ where σ_y is the material yield stress.

Therefore, in the presence of bending stress σ_b, the plastic moment of resistance of the flange is reduced to $M_f[1 - (\sigma_b/\sigma_f)^2]$. It is assumed also that the plastic moment of resistance of the web is reduced to $M_w[1 - (\sigma_b/\sigma_w)^2]$, even though the coexistent bending stress acts perpendicular to the plane of bending. Applying the correction factor $[1 - (\sigma_b/\sigma_w)^2]$ to the plastic moment of resistance is equivalent to reducing the web yield stress by the same factor. Hence, if σ_w in eqn (3.28) is reduced to $\sigma_w[1 - (\sigma_b/\sigma_w)^2]$, the collapse load is given by

$$P_u = 0.5t_w^2 \left\{ E\sigma_w \frac{t_f}{t_w} \right\}^{0.5} \left\{ 1 + \frac{3c}{d} \left(\frac{t_w}{t_f} \right)^{1.5} \right\} \left\{ 1 - \left(\frac{\sigma_b}{\sigma_w} \right)^2 \right\}^{0.5} \quad (3.32)$$

It is recommended that the value of P_u given by eqn (3.31) should be reduced by the same factor.

3.5.4 Mechanism Solutions for Distributed Edge Loading

Modified versions of the mechanism solutions already discussed were proposed by Roberts and Chong (1981) for situations where the edge loading is uniformly distributed between vertical web stiffeners. The loading and assumed mechanism of collapse are shown in Fig. 3.12. It is

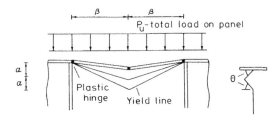

FIG. 3.12.   Modified collapse mechanism—web bending.

assumed that three plastic hinges form in the flange—two adjacent to and one mid-way between the vertical web stiffeners.

If at the collapse load the central plastic hinge in the flange is assumed to move vertically through a small distance δv, the rotation of the plastic hinges in the flange is $\delta v/\beta$ (twice this value for the central plastic hinge). The corresponding rotation of the yield lines in the web is $\delta v/2\alpha \cos \theta$ (twice this value along the central yield line). Hence, equating the work done by the applied load to the internal dissipation of plastic energy gives

$$P_u = \frac{8M_f}{\beta} + \frac{8\beta M_w}{\alpha \cos \theta} \quad (3.33)$$

An estimate of the deflection of the flange just prior to collapse can be made using elastic theory as in Section 3.5.1. This leads to an equation which is identical to eqn (3.20) which can be rearranged as

$$\sin \theta = 1 - \frac{M_f \beta^2}{12EI_f \alpha} \tag{3.34}$$

The uniformly distributed edge loading produces compressive membrane stresses σ_m in the web which reduce the plastic moment of resistance of the web to $M_w[1 - (\sigma_m/\sigma_w)^2]$. If σ_m is assumed uniform, then σ_m can be deduced from the web contribution to the collapse load, i.e. the second term on the right-hand side of eqn (3.34). Hence

$$\sigma_m = \frac{4M_w}{\alpha \cos \theta t_w} \tag{3.35}$$

and introducing the reduced plastic moment of the web into eqn (3.34) gives, as a first approximation to the collapse load

$$P_u = \frac{8M_f}{\beta} + \frac{8\beta M_w}{\alpha \cos \theta} \left\{ 1 - \left(\frac{4M_w}{\alpha \cos \theta t_w \sigma_w} \right)^2 \right\} \tag{3.36}$$

If α is taken again simply as

$$\alpha = 25 t_w \tag{3.37}$$

eqns (3.34) and (3.36) and the condition that $\beta = b/2$ lead to a value for the collapse load.

The possibility of failure being initiated by direct yielding of the web can be analysed by considering the failure mechanism shown in Fig. 3.13. Equating the work done by the applied load to the internal dissipation of plastic energy gives

$$P_u = \frac{8M_f}{\beta} + 2\beta \sigma_w t_w \tag{3.38}$$

The collapse load is taken as the lesser of the values given by eqns (3.36) and (3.38). As for partial edge loading, an allowance for the influence of

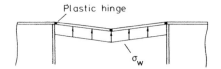

FIG. 3.13.   Modified collapse mechanism—web yielding.

coexistent bending stresses σ_b can be made by reducing the collapse load by a factor

$$\left\{1 - \left(\frac{\sigma_b}{\sigma_w}\right)^2\right\}^{0\cdot5} \qquad (3.39)$$

3.6 COMPARISON WITH TEST DATA

The values of the collapse loads predicted in accordance with Sections 3.5.1, 3.5.2 and 3.5.3 have been compared with over one hundred test results from various sources, on girders having web thickness varying from 1 to 5 mm (Roberts, 1981a). It was found that in all cases, eqn (3.32), which was derived from the mechanism for web bending, gave the lowest value of P_u. A lower limit of three was imposed on the value of t_f/t_w used in eqn (3.32) since this equation tends to underestimate the collapse load of girders with very thin webs and having $t_f/t_w < 3$. The results of this comparison are shown in Fig. 3.14 in which P_{ex} and P_{pr} are the experimental and predicted values of the collapse load respectively. For these results the mean value of the ratio P_{ex}/P_{pr} is 1·43 and the coefficient of variation is 15·8 %.

To illustrate the relationship between the collapse load and the elastic critical load of an assumed simply-supported web panel, the same results

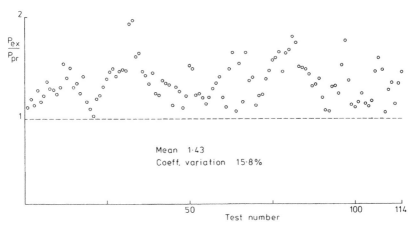

FIG. 3.14. Comparison of predicted collapse loads with test data.

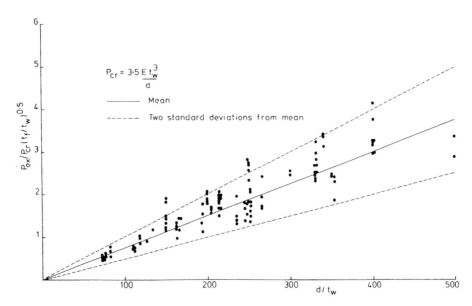

FIG. 3.15.   Comparison of test data with elastic critical loads.

are plotted in a different form in Fig. 3.15. P_{cr} is the elastic critical load, which for simplicity was taken as

$$P_{cr} = 3 \cdot 5 \, \frac{\pi^2 D}{d} \tag{3.40}$$

this being the value for a web panel having $b/d = 1$ and $c/d = 0 \cdot 25$. The continuous-line in Fig. 3.15 represents the predicted value of the collapse load given by eqn (3.32) multiplied by the mean value of P_{ex}/P_{pr}. It therefore represents the mean of the results shown in Fig. 3.15. The dashed-lines are spaced at two standard deviations on each side of the mean and therefore represent the 95 % confidence limits. This plot indicates that the collapse load and elastic critical load are approximately equal for girders having $b/d = 1$, $t_f/t_w = 3$ and $d/t_w = 75$. For girders having $d/t_w > 75$, the collapse load is significantly greater than the elastic critical load. Unfortunately, there is little test data available for girders having web thicknesses greater than 5 mm and $d/t_w < 75$.

The values of the collapse loads predicted in accordance with Section 3.5.4 for uniformly distributed edge loading have been compared by Roberts and Chong (1981) with a limited number of tests, carried out by

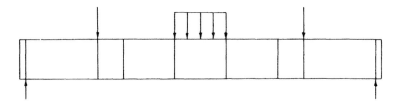

FIG. 3.16.   Loading conditions for uniformly-distributed edge loading.

Bossert and Ostapenko (1967). These tests were performed on medium-span girders with vertical web stiffeners—the web depth and thickness being 915 mm and 3 mm respectively. A single web panel was loaded with a uniformly-distributed edge load and additional concentrated loads were applied directly to the vertical web stiffeners to vary the magnitude of the coexistent bending stress. The type of girder tested and loading arrangement are shown in Fig. 3.16 and the values of P_{ex}/P_{pr} are plotted in Fig. 3.17. The predicted value for the test in which the distributed load was applied via a wooden beam is somewhat low since the wooden beam transferred some of the load directly to the vertical web stiffeners.

Having shown that the mechanism solution based on web bending provides a satisfactory prediction of the collapse load of plate girders with thin webs (less than 5 mm) subjected to patch loading and distributed edge loading, it is of interest to investigate whether or not the mechanism

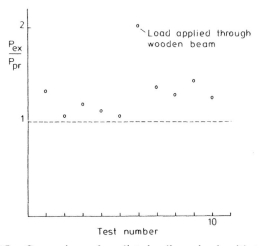

FIG. 3.17.   Comparison of predicted collapse loads with test data.

solutions based on web yielding have any significance. A consideration of the test data for girders E10-10-1/1 and E10-10-2/1 given in Table 3.1 leads to interesting conclusions.

The collapse load for web bending given by eqn (3.28) is denoted by P_{ub} and for web yielding given by eqn (3.31) is denoted by P_{uy}. Values of P_{ex}, P_{uy} and P_{ub} are given in Table 3.2 and for both tests, P_{uy} is less than P_{ub}.

TABLE 3.2

Girder	P_{ex} (kN)	P_{uy} (kN)	P_{ub} (kN)
E10-10-1/1	716	179	585
E10-10-2/1	787	438	688

Inspection of the load–deflection curves shown in Fig. 3.4 reveals a change in slope at loads not too much greater than P_{uy}; this change in slope being due to significant membrane yielding of the web. The girders sustained, however, significantly higher loads than P_{uy} and eventually failed in the familiar crippling mode at loads above P_{ub}. It appears therefore that P_{ub} provides a satisfactory prediction of the collapse load for girders having thick as well as thin webs but that at loads above P_{uy}, significant membrane yielding of the web may occur. Further comparisons made by Roberts (1981b) with tests on rolled I-beams having web thicknesses up to 18 mm and d/t_w ratios as low as 20, compressed by two equal and opposite concentrated forces, confirm these conclusions.

3.7   SUMMARY AND CONCLUSIONS

The collapse or crippling of plate girders subjected to patch loading and distributed edge loading is a complex problem involving both material and geometric nonlinearity. Solutions are now available for the elastic critical loads of idealised web panels subjected to a variety of combined loading but such solutions show little or no correlation with experimental collapse loads. To date no rigorous solutions are available for predicting the collapse of plate girders subjected to edge loading but the simple approximate solutions presented herein and elsewhere enable designers to predict collapse with an accuracy that is acceptable for practical purposes. It is also possible to predict the loads above which membrane yielding of the web becomes pronounced.

REFERENCES

ALFUTOV, N. A. and BALABUKH, L. I. (1967) On the possibility of solving plate stability problems without a preliminary determination of the initial state of stress. *PMM*, **31**(4), 730–6.

ALFUTOV, N. A. and BALABUKH, L. I. (1968) Energy criterion of the stability of elastic bodies which does not require the determination of the initial stress–strain state. *PMM*, **32**(4), 726–31.

BERGFELT, A. (1971) Studies and tests on slender plate girders without stiffeners—shear strength and local web crippling. *IABSE Colloquium on Design of Plate and Box Girders for Ultimate Strength*, London.

BERGFELT, A. (1979) Patch loading on a slender web. Chalmers University of Technology, Goteborg, Inst. For Konst., Stal-Och Trabyggnad, Int. skr. S79:1.

BERGFELT, A. and HOVIK, J. (1968) Thin walled deep plate girders under static loads. *Proc. 8th Congr. IABSE*, New York.

BERGFELT, A. and HOVIK, J. (1970) Shear failure and local web crippling in thin walled plate girders. Chalmers University of Technology, Goteborg, Inst. For Konst., Stal-Och Trabyggnad, Int. skr. S70:11b.

BERGFELT, A. and LINDGREN, S. (1974) Local web crippling in thin walled plate girders under concentrated loads. Chalmers University of Technology, Goteborg, Inst. For Konst. Stal-Och Trabyggnad, Int. skr. S74:5.

BOSSERT, T. W. and OSTAPENKO, A. (1967) Buckling and ultimate loads for plate girder web plates under edge loading. Report No. 319.1, Fritz Engineering Laboratory, Lehigh University, Bethlehem.

DRDACKY, M. and NOVOTNY, R. (1977) Partial edge load carrying capacity tests on thin plate girder webs. *Acta Tech.*, Prague, No. 5.

DUBAS, P. and GEHRI, E. (1975) Behaviour of webs under concentrated loads acting between vertical stiffeners. *ECCS Commission 8.3*, Zurich.

GIRKMANN, K. (1936) Int. Assoc. for Bridge and Struct. Eng. Final Report.

GRANHOLM, C. A. (1976) Light girders. Girders with slender flanges and web (Original reports in Swedish—summaries in English by A. Bergfelt). Chalmers University of Technology, Goteborg, Inst. For Konst., Stal-Och Trabyggnad, Int. skr. S76:14.

KHAN, M. Z. and JOHNS, K. C. (1975) Buckling of web plates under combined loadings. *Proc. ASCE*, ST10, 2079–92.

KHAN, M. Z., JOHNS, K. C. and HAYMAN, R. (1977) Buckling of plates with partially loaded edges. *Proc. ASCE*, ST3, 547–58.

KHAN, M. Z. and WALKER, A. C. (1972) Buckling of plates subjected to localised edge loading. *Structural Engineer*, **50**(6), 225–32.

LEGGETT, D. M. A. (1937) The effect of two isolated forces on the elastic stability of a flat rectangular plate. *Proc. Cambridge Phil. Soc.*, **33**, 325–39.

ROBERTS, T. M. (1980) Experimental and theoretical studies on slender plate girders subjected to edge loading. University College, Cardiff, Dept. Civil and Struct. Eng. Report.

ROBERTS, T. M. (1981a) Slender plate girders subjected to edge loading. *Proc. Inst. Civ. Engrs*, Part 2, 71, 805–19.

ROBERTS, T. M. (1981b) Stocky and stiffened plate girders subjected to edge loading. University College, Cardiff, Dept. Civil and Struct. Eng. Report.

ROBERTS, T. M. and CHONG, C. K. (1981) Collapse of plate girders under edge loading. *Proc. ASCE*, ST8, 1503–9.

ROBERTS, T. M. and ROCKEY, K. C. (1977) A method for predicting the collapse load of a plate girder when subjected to patch loading in the plane of the web. University College, Cardiff, Dept. Civil and Struct. Eng. Report.

ROBERTS, T. M. and ROCKEY, K. C. (1979) A mechanism solution for predicting the collapse loads of slender plate girders when subjected to in-plane patch loading. *Proc. Inst. Civ. Engrs*, Part 2, 67, 155–75.

ROCKEY, K. C. and BAGCHI, D. K. (1970) Buckling of plate girder webs under partial edge loadings. *Int. J. Mech. Sci.*, **12**, 61–76.

ŠKALOUD, M. and DRDACKY, M. (1975) Ultimate load design of webs of steel plate girders, Part 3. Webs under concentrated loads. Staveb Cas. 23 c 3 VEDA, Bratislava.

ŠKALOUD, M. and NOVAK, P. (1975) Post buckled behaviour of webs under partial edge loading. Acad. Sci. Rep., 85, No. 3, Prague.

SOMMERFIELD, A. Z. (1906) *Z. Math. Phys.*, **54**, 113.

TIMOSHENKO, S. P. (1910) *Z. Math. Phys.*, **58**, 357.

TIMOSHENKO, S. P. and GERE, J. M. (1961) *Theory of Elastic Stability*, 2nd Edition, McGraw-Hill, Tokyo.

WHITE, R. M. and COTTINGHAM, W. (1962) Stability of plates under partial edge loadings. *Proc. ASCE*, **88**, 67–86.

ZETLIN, L. (1955) Elastic instability of flat plates subjected to partial edge loads. *Proc. ASCE*, **81**, Paper 795, 1–24.

Chapter 4

OPTIMUM RIGIDITY OF STIFFENERS OF WEBS AND FLANGES

M. Škaloud

*Institute of Theoretical and Applied Mechanics,
Czechoslovak Academy of Sciences, Prague, Czechoslovakia*

SUMMARY

The optimum rigidity of stiffeners is expressed as a product of (i) the quantity γ^*, resulting from linear theory of web buckling, and (ii) a correction factor k, which expresses the effect of initial imperfections, and of post-buckled behaviour. Formulae for γ^* for the most frequently encountered cases of stiffening and loading are presented, and the magnitude of the factor k is discussed in the light of theoretical and experimental evidence.

4.1 TWO CONCEPTS OF DESIGN OF STIFFENED WEBS

Two concepts are possible in design of the stiffened webs and flanges of steel structures, viz.

(i)   the concept of rigid stiffeners,
(ii)   that of flexible stiffeners.

In the first approach all stiffeners are proportioned so as to provide the web sheet with rigid support, thereby inducing in the stiffeners nodal lines of the buckled surface of the web. With this approach, one obtains the best possible boundary conditions for the buckling of the web sheet and, consequently, the highest possible ultimate load of the stiffened web. Thus,

103

one can arrive at the minimum thickness of the web sheet, of course at the expense of putting a sufficient amount of material into the stiffening.

Also, in the second approach, the presence of flexible ribs improves the behaviour of webs, more or less so in terms of the size of the ribs, but they are not rigid enough to provide a sufficiently rigid support for the web sheet. Therefore, the stiffeners deflect with the buckling sheet. Then the ultimate load of the stiffened web cannot be as high as in the case of rigid stiffeners; this leads to larger thicknesses of the web sheet, but on the other hand, the amount of material put into the ribs is also less.

It can be the aim of an optimisation study to choose the more economic alternative in the particular case, and eventually the final choice is left to the designer. But, by and large, the concept of rigid ribs is employed much more frequently than the other approach. Its popularity is due not only to economic reasons but also to the fact that the analysis involved is simpler and much less time-consuming; once the stiffeners are designed as rigid, the analysis of the stiffened web as a whole can be reduced to an analysis of the individual sheet panels between the stiffeners. For this reason in this chapter we are going to deal with the concept of rigid ribs thoroughly.

4.2 THE NOTION OF THE OPTIMUM RIGIDITY OF STIFFENERS

Let us measure the supporting capability of stiffeners by their relative flexural rigidity

$$\gamma = \frac{EI_s}{Db}$$

where E = Young's modulus, I_s = moment of inertia of the stiffener, $D = Et^3/12(1 - v^2)$ = flexural rigidity of unit width of web (t denotes web thickness and v Poisson's ratio of web material), b = web depth.

The following question is of fundamental importance:

What is the minimum value of the relative flexural rigidity, γ, which ensures that the stiffener under consideration would behave as a rigid one?

The corresponding value of γ is called the optimum rigidity of the stiffener.

It is obvious that the reply to the question depends very significantly on the method of the analysis. Two courses are open: either we can start from the assumption of an 'ideal' web without initial imperfections and base our analysis on linear buckling theory, or we can make allowance for the presence of unavoidable initial imperfections and take account of the actual ultimate load behaviour of the web.

4.3   THE OPTIMUM RIGIDITY OF STIFFENERS USING THE LINEAR BUCKLING THEORY OF 'IDEAL' WEBS

4.3.1 Definition of the Optimum Rigidity γ^* of Stiffeners Using the Linear Buckling Theory

The linear buckling theory is based on two assumptions:

(i)   The first is the assumption of 'ideal' webs, according to which webs are perfectly plane prior to loading, with load applied exactly in the middle plane of the web, with no effect of residual stresses.

(ii)   Further, it is assumed that web deflections are very small compared to web thickness, so that the linear theory describes web buckling with sufficient accuracy.

The limit state of the web is then given by bifurcation of equilibrium (Fig. 4.1), the corresponding value of load, p_{cr}, being called 'critical'. The design of webs is then based on the idea that a web remains perfectly plane as long as the load is inferior to the critical value and that the web buckles (and collapses) as soon as the critical load has been exceeded. Hence, the limit load of the web is supposed to equal its critical loading.

When the web under consideration is fitted with ribs, the assumption of 'ideal' stiffeners is also introduced in the analysis, i.e. only perfectly straight stiffeners, without any residual stresses and other material or geometrical imperfections are considered.

As the design of webs is based on the critical load, the design of stiffeners is governed by the requirement that they be rigid and fully effective under this critical load. And the minimum value of the flexural rigidity γ that makes it possible for the stiffener to achieve this objective is called 'optimum' and denoted as γ^*.

When quantifying the requirement that the stiffener be 'fully effective and rigid' under the critical loading, it is seen that a unique definition—applicable in all cases—is not used in practice. However, we can always employ one of the following three definitions:

1. If the stiffener is in a possible nodal line of the corresponding unstiffened web, there can exist a finite flexural rigidity of the stiffener at which the critical load p_{cr} attains its maximum possible value $p_{cr\,max}$. The

M. ŠKALOUD

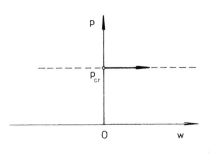

a

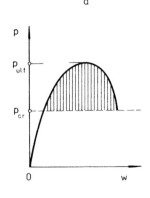

b

FIG. 4.1.   (a) Behaviour of an 'ideal' web; (b) behaviour of an 'actual' web.

optimum rigidity then follows from the condition of existence of two equilibrium states of the rib: i.e. an undeformed state and a deformed one. For example, for a uniformly compressed plate reinforced on the central line by a longitudinal rib of flexural rigidity $\gamma = \gamma^*$ (Fig. 4.2) two stiffener deformations are possible, one being a symmetric configuration with deflected stiffener, the other an antisymmetric configuration where the stiffener remains straight.

The relationship between the critical stress σ_{cr} and the flexural rigidity γ is shown in Fig. 4.3. The optimum rigidity defined in this way is termed 'optimum rigidity of the first kind' and denoted as γ_1^*. It ensures that the stiffener remains perfectly rigid when the adjacent plate panels buckle, and determines its most economic dimensions. A further increase in rib stiffness would not bring about any increase in the critical load.

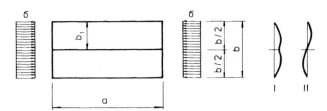

FIG. 4.2. A compression plate stiffened by one longitudinal rib on the central line.

To conclude this comment on the optimum rigidity of the first kind, it remains to be explained when exactly this rigidity can be obtained. As has been mentioned above, such a value can be encountered when the central line of the stiffener coincides with a possible nodal line of the corresponding unstiffened plate. This is understandable in view of the fact that in the first case the stiffener is loaded by no reactions of the buckling plate and therefore is not forced to deflect. This condition is, however, not the only one. Moreover, it is necessary that the critical stress $p_{cr}(\gamma_1^*)$, related to the value γ_1^*, should be less than the critical load p_{cr} $(\gamma = \infty)$, which corresponds to an infinitely rigid stiffener. Unless this is the case, the critical load $p_{cr}(\gamma_1^*)$ can never be attained, and the optimum rigidity γ_1^* has no practical value.

2. The optimum rigidity of the first kind exists only for some positions of the stiffener (or, more accurately for some types of stiffening) and only for some loadings. In general, there exists no finite rigidity that would guarantee that the rib remains straight and the critical load reaches its maximum possible value. The buckling load attains its maximum for an infinitely rigid stiffener. Nevertheless, even in this case a flexural rigidity can sometimes be found for which two equilibrium states of the rib are possible. In this case both are deformed. For example, for a uniformly compressed

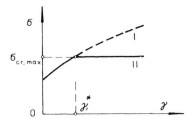

FIG. 4.3. Optimum rigidity of the first kind.

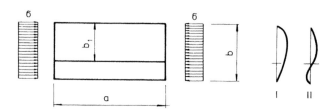

FIG. 4.4. A compression plate stiffened by one longitudinal rib off the central line.

plate, reinforced by an eccentrically located longitudinal stiffener (Fig. 4.4) of rigidity $\gamma = \gamma_{II}^*$, two equilibrium states I and II are possible. The relationship γ versus σ_{cr} is plotted in Fig. 4.5. The optimum rigidity defined in this way is called 'optimum rigidity of the second kind' and designated as γ_{II}^*. Unlike the value γ_I^*, it does not guarantee that a nodal line forms along the axis of the rib. The rib slightly deflects. By further enlarging the rigidity of the stiffener beyond the value γ_{II}^*, a further increase—with respect to $\sigma_{cr\,M}$—in the critical load is obtained; but in most cases it is slight. The maximum possible critical load, $\sigma_{cr\,max}$, is attained only for $\gamma \to \infty$.

3. In most practical cases neither the optimum rigidity of the first kind, nor that of the second kind exists. Nonetheless, even in these cases it is advantageous to have a suitable basis for stiffener design. The optimum rigidity is then defined as that value of γ at which the critical load of the whole stiffened web is equal to the critical stress of the most unfavourably loaded and most slender web panel provided this panel is assumed to be simply supported on all four boundaries. Such a flexural rigidity is called 'optimum rigidity of the third kind' and is denoted as γ_{III}^*.

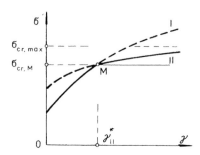

FIG. 4.5. Optimum rigidity of the second kind.

This definition is simple and applicable for all sorts of stiffening and loading of webs. When γ_I^* exists, always $\gamma_{III}^* = \gamma_I^*$. Despite the fact that the optimum rigidity of the third kind does not guarantee that a nodal line forms at the stiffener axis and that the critical load attains the highest possible value, it significantly simplifies the design of stiffened webs. Their analysis is then reduced to the design of individual web panels, which (as already stated in Section 4.1) is one of the reasons for the popularity of the concept of rigid ribs. That is why the optimum rigidity of the third kind forms the basis of the rules for the design of stiffened webs and flanges. It is simply called 'optimum rigidity' and designated as γ^*. Most of literature dealing with the behaviour of stiffened webs has been devoted to this approach to the optimum rigidity of stiffeners.

4.3.2 The Values of γ^* for Different Kinds of Stiffening and Loading of Webs

Numerous studies to date have been concerned with determination of the optimum rigidity γ^* for various kinds of stiffening and loading of webs and flange plates. Probably most comprehensive are the two well-known books by Klöppel and his associates (Klöppel and Scheer, 1960; Klöppel and Möller, 1968). These and other stability monographs give not only a detailed description of the solution to the problem, but also a good many formulae and charts for γ^*.

One of the problems which remained unsolved until recently was that of the optimum rigidity of the ribs of longitudinally stiffened compression flanges—the stability literature gives values merely for compressed plates fitted with one and two (and partially with three) stiffeners. A solution to this problem, contemplating the flange as a discretely stiffened plate, was presented by Škaloud and Zörnerová (1977), where charts for γ^*-values were given for compression flanges stiffened by 1–20 longitudinal ribs.

The effect of the configuration of the ribs upon the buckling of longitudinally stiffened compression flanges was studied by Křístek and Škaloud (1977) and Škaloud and Křístek (1977, 1981) by means of folded plate theory. A very pronounced favourable influence of closed-section trough stiffeners follows from this investigation.

To conclude this section, the formulae for γ^* for the most frequently encountered cases of stiffened webs and flanges are listed in Tables 4.1(a)–(c). Most of them have been taken from available stability literature; the formulae in the third row of Table 4.1(a) summarises with sufficient accuracy the results obtained by Zörnerová and Škaloud.

M. ŠKALOUD

TABLE 4.1(a)

Kind of stiffening and loading	Range of validity	γ^*
	$\alpha < \sqrt{(8(1+2\delta)) - 1}$	$\gamma^* = (0{\cdot}53 + 0{\cdot}47\psi)\left\{\dfrac{\alpha^2}{2}[16(1+2\delta) - 2] - \dfrac{\alpha^4}{2} + \dfrac{1+2\delta}{2}\right\}$
	$\alpha > \sqrt{(8(1+2\delta)) - 1}$	$\gamma^* = (0{\cdot}53 + 0{\cdot}47\psi)\left\{\dfrac{1}{2}[8(1+2\delta) - 1]^2 + \dfrac{1+2\delta}{2}\right\}$
	$\alpha < \sqrt{(18(1+3\delta)) - 1}$	$\gamma^* = \dfrac{\alpha^2}{3}[36(1+3\delta) - 2] - \dfrac{\alpha^4}{3} + \dfrac{1+3\delta}{3}$
	$\alpha > \sqrt{(18(1+3\delta)) - 1}$	$\gamma^* = \tfrac{1}{3}[18(1+3\delta) - 1]^2 + \dfrac{1+3\delta}{3}$
	$\alpha < n\sqrt{(2(1+n\delta))}$	$\gamma^* = \dfrac{1}{n}[4n^2(1+n\delta)\alpha^2 - \alpha^4]$
	$\alpha > n\sqrt{(2(1+n\delta))}$	$\gamma^* = 4n^3(1+n\delta)^2$

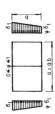

$0.4 \leq \alpha \leq 1.4$

$$\gamma^* = \frac{4\left(\dfrac{4}{\alpha^2} - \dfrac{\alpha}{4}\right)}{\pi^2 \alpha \left(1 - \dfrac{\pi^2 \alpha^4}{12a - 48}\right)}$$

$\alpha > 1.4$

practically ineffective

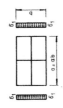

$0.9 \leq \alpha \leq 1.1$

$$\gamma_l^* = \frac{(1+\alpha^2)^2[(1+2\delta_l) - 1]}{2(1+p\alpha^3)}$$

where

$$p = \frac{\gamma_t}{\gamma_l} = \frac{I_t}{I_l}$$

subscript l is related to longitudinal stiffener

subscript t to transverse stiffener

TABLE 4.1(b)

Kind of stiffening and loading	Range of validity	γ^*
	$\alpha \leq 0.5$	$\gamma^* = 1.3$
	$\alpha > 0.5$	$\gamma^* = 2.4 + 18.4\delta$ $\left\{ \begin{array}{l} \gamma^* = (12 + 92\delta)(\alpha - 0.3) \\ \text{with the maximum value} \\ \gamma^* = 16 + 200\delta \end{array} \right.$
	$0.5 \leq \alpha \leq 1.5$	$\gamma = 3.87 + 5.1\alpha + (8.82 + 77.6\delta)\alpha^2$
	$0.6 \leq \alpha \leq 0.935$ $\alpha > 0.935$	$\gamma^* = 6.2 - 12.7\alpha + 6.5\alpha^2$ practically ineffective

TABLE 4.1(c)

Kind of stiffening and loading	Range of validity	γ^*
	$0.5 \leq \alpha \leq 2.0$	$\gamma^* = 5.4\alpha^2(2\alpha + 2.5\alpha^2 - \alpha^3 - 1)$
	$0.3 \leq \alpha \leq 1.0$	$\gamma^* = 12.1\alpha^2(4.4\alpha - 1)$
	$0.5 \leq \alpha \leq 2.0$	$\gamma^* = 7.2\alpha^2(1 - 3.3\alpha + 3.9\alpha^2 - 1.1\alpha^3)$
	$0.5 \leq \alpha \leq 2.0$	$\gamma^* = \dfrac{5.4}{\alpha}\left(\dfrac{2}{\alpha} + \dfrac{2.5}{\alpha^2} - \dfrac{1}{\alpha^3} - 1\right)$
	$1.0 \leq \alpha \leq 3.3$	$\gamma^* = \dfrac{12.1}{\alpha}\left(\dfrac{4.4}{\alpha} - 1\right)$
	$0.5 \leq \alpha \leq 2.0$	$\gamma^* = \dfrac{7.2}{\alpha}\left(1 - \dfrac{3.3}{\alpha} - \dfrac{3.9}{\alpha^2} - \dfrac{1.1}{\alpha^3}\right)$
	$0.2 \leq \alpha \leq 1.0$	$\gamma^* = \dfrac{28}{\alpha^2} - 20$

4.4 THE OPTIMUM RIGIDITY OF STIFFENERS CONSIDERING THE ULTIMATE LOAD BEHAVIOUR OF INITIALLY IMPERFECT WEBS

4.4.1 Definition of the Optimum Rigidity of Stiffeners Considering the Ultimate Load Behaviour

The linear theory approach to the problem of the optimum rigidity of stiffeners, presented in Section 4.3, is based on unrealistic assumptions. Hence, it is not able to describe the actual behaviour of stiffened webs and flanges with sufficient accuracy.

To begin with, the webs and flanges of ordinary steel structures do not accord with the simplified model of an 'ideal' plate element as described above. The webs and flanges of all constructed girders, columns, etc., exhibit initial irregularities (initial curvature, eccentricity, residual stresses, etc.). Consequently, they deform from the very beginning of the loading process, and in most cases the deflections grow monotonously without any sudden change in their neighbourhood of the critical load (Fig. 4.1(a)). Similar unavoidable imperfections are encountered in stiffeners.

The web (and flange) deflections cannot be regarded as small, but they are of the order of web thickness. A new stress components, sc membrane stresses, uniformly distributed over the web thickness develop in the web sheet. These stresses, which are not accounted for by the linear theory of web buckling, have a pronounced beneficial effect on the behaviour of plated structures. They substantially slow down the increase in web deflections and stresses so that the load-carrying capacity of webs, p_{ult}, can be (with the exception of the web plates of low depth-to-thickness ratios) considerably larger than the critical load p_{cr} (in the case of very slender webs subjected to combined shear and bending even several times).

Then the optimum rigidity, this time denoted as γ_0, is defined as: *the minimum value of the relative flexural rigidity, γ, of an initially imperfect stiffener on an initially imperfect web (or flange) which ensures that the stiffener under consideration remains (at least practically) rigid in the whole post-buckled range of the web behaviour, i.e. up to its ultimate load.*

To the author's knowledge, it was Massonnet who first paid attention to this problem in his experimental studies (Massonnet, 1954; Massonnet *et al.*, 1962). He demonstrated that stiffeners proportioned by means of the γ^*-values deflected with the web in the post-buckled range. He also noted that for the stiffeners on his test girders to remain straight and fully effective up to the collapse of the girder, it was necessary that they should possess a flexural rigidity equal to several times γ^*. As this observation is of

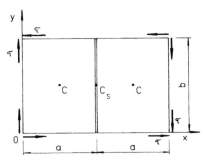

FIG. 4.6. A web subjected to shear and stiffened by one central transverse stiffener.

fundamental importance, further studies, both theoretical and experimental, were needed for its verification.

4.4.2 Theoretical Investigation into the Post-Buckled Behaviour of Stiffened Webs

The post-critical performance of stiffened webs is demonstrated on the example of a rectangular web of length $2a$ and depth b, subjected to a uniform shear, τ, and reinforced by a central vertical stiffener (Fig. 4.6). The coordinate system (x, y) is also indicated in Fig. 4.6.

The web is supposed to be fabricated from an ideally elasto-plastic steel whose behaviour is governed by Prandtl's stress–strain relationship. Further, it is assumed that the behaviour of the web is perfectly elastic in the whole range of the performance of the web studied. The investigation is limited to studying the effect of the flexural rigidity of the rib upon the post-critical behaviour of the web. It is accordingly assumed that the stiffener is such that all torsional phenomena of the stiffener can be disregarded.

The investigation was carried out by Massonnet et al. (1968), Škaloud et al. (1967) and Škaloud (1970). A brief description of the solution is presented below.

If an initial curvature and residual stresses are not considered, the differential equations governing the problem are as follows:

$$\frac{\partial^4 \Phi}{\partial x^4} + 2 \frac{\partial^4 \Phi}{\partial x^2 y^2} + \frac{\partial^4 \Phi}{\partial y^4} = E \left[\left(\frac{\partial^2 w}{\partial x \, \partial y} \right) - \frac{\partial^2 w}{\partial x^2} \frac{\partial^2 w}{\partial y^2} \right] \qquad (4.1a)$$

$$\frac{D}{t} \left(\frac{\partial^4 w}{\partial x^4} + 2 \frac{\partial^4 w}{\partial x^2 y^2} + \frac{\partial^4 w}{\partial y^4} \right) = \frac{\partial^2 \Phi}{\partial y^2} \frac{\partial^2 w}{\partial x^2} + \frac{\partial^2 \Phi}{\partial x^2} \frac{\partial^2 w}{\partial y^2} - \frac{\partial^2 \Phi}{\partial x \, \partial y} \frac{\partial^2 w}{\partial x \, \partial y}$$

$$(4.1b)$$

The boundary conditions are determined by the support of the web.

It will be assumed that the web is simply supported on all four edges, from which the following boundary equations of the web deflections follow:

(a) for $x = 0$, $x = 2a$: $\quad w = 0$ $\hspace{4cm}$ (4.2a)

$\quad\quad$ for $y = 0$, $y = b$: $\quad w = 0$ $\hspace{4cm}$ (4.2b)

(b) for $x = 0$, $x = 2a$: $\quad M_x = \dfrac{\partial^2 w}{\partial x^2} + v\,\dfrac{\partial^2 w}{\partial y^2} = 0$ $\hspace{2cm}$ (4.3a)

$\quad\quad$ for $y = 0$, $y = b$: $\quad M_y = \dfrac{\partial^2 w}{\partial y^2} + v\,\dfrac{\partial^2 w}{\partial x^2} = 0$ $\hspace{2cm}$ (4.3b)

The boundary conditions for the stress function Φ depend on the rigidity of the peripheral elements of the web. Two cases will be considered:

(A) The flexural rigidity of the peripheral elements is such that the web edges remain straight:

for $x = 0$, $x = 2a$:

$$e_x = \frac{1}{2a} \int_0^{2a} \frac{\partial u}{\partial x}\,\mathrm{d}x = \frac{1}{2a}\int_0^{2a} \left[\frac{1}{E}\left(\frac{\partial^2 \Phi}{\partial y^2} - v\,\frac{\partial^2 \Phi}{\partial x^2} \right) - \frac{1}{2}\left(\frac{\partial w}{\partial x} \right)^2 \right]\mathrm{d}x = \text{constant}$$

for $y = 0$, $y = b$:

$$e_x = \frac{1}{b} \int_0^{b} \frac{\partial v}{\partial y}\,\mathrm{d}y = \frac{1}{b}\int_0^{b} \left[\frac{1}{E}\left(\frac{\partial^2 \Phi}{\partial x^2} - v\,\frac{\partial^2 \Phi}{\partial y^2} \right) - \frac{1}{2}\left(\frac{\partial w}{\partial y} \right)^2 \right]\mathrm{d}y = \text{constant}$$

Here again two alternatives can be envisaged:

(A$_1$) The opposite web edges remain straight and do not approach:

for $\hspace{3.5cm}$ $x = 0$, $x = 2a$: $\hspace{1cm}$ $e_x = 0$ $\hspace{2.5cm}$ (4.4a)

for $\hspace{3.5cm}$ $y = 0$, $y = b$: $\hspace{1cm}$ $e_y = 0$ $\hspace{2.5cm}$ (4.4b)

(A$_2$) The opposite web edges, remaining straight, can move freely in plane:

for $\hspace{3.5cm}$ $x = 0$, $x = 2a$: $\hspace{1cm}$ $p_x = 0$ $\hspace{2.5cm}$ (4.5a)

for $\hspace{3.5cm}$ $y = 0$, $y = b$: $\hspace{1cm}$ $p_y = 0$ $\hspace{2.5cm}$ (4.5b)

where p_x and p_y denote the average values of the stresses acting on the edges of the web.

(B) The peripheral elements of the web are perfectly flexible:

for $\qquad x = 0, \; x = 2a: \qquad \dfrac{\partial^2 \Phi}{\partial y^2} = 0 \qquad$ (4.6a)

for $\qquad y = 0, \; y = b: \qquad \dfrac{\partial^2 \Phi}{\partial x^2} = 0 \qquad$ (4.6b)

Besides condition A or condition B the function Φ must satisfy the following equations:

for $\qquad x = 0, \; x = 2a: \qquad -\dfrac{\partial^2 \Phi}{\partial x \, \partial y} = \tau \qquad$ (4.7a)

for $\qquad x = 0, \; y = b: \qquad -\dfrac{\partial^2 \Phi}{\partial x \, \partial y} = \tau \qquad$ (4.7b)

The choice of an assumed buckled surface is discussed below.

Out of permissible functions w_{ij} having all continuous derivatives up to the 4th-order and satisfying boundary conditions (4.2) and (4.3), the following coordinate functions are selected:

$$w_{ij} = \sin \frac{i\pi x}{2a} \sin \frac{j\pi y}{b} \qquad i, j = 1, 2, \ldots$$

If the web length $2a$ is not very different from the depth b, the buckled pattern of an unstiffened web consists of a large buckling half-wave following the tension diagonal and of two smaller half-waves, each of them situated near one end of the tension diagonal and on opposite sides of the predominant buckling half-wave.

Therefore, an assumption for the deflection surface, w, must be chosen that for small stiffener rigidities would be compatible with the buckled pattern discussed above. If, on the contrary, the stiffener is very rigid, dividing the web into two separate panels of length a, each of these panels behaves as the whole web similar to the case of an unstiffened web (considered above). For intermediate stiffener rigidities the shape of the buckled pattern will be between the two limiting cases discussed above.

Hence, the assumption for the deflection surface, w, must satisfy the requirements discussed in this section. Furthermore, in view of the fact that solutions to problems of nonlinear theory of large deflections are very complex and time-consuming, it is necessary to limit oneself to a reasonable number of the coordinate functions w_{ij}. Six functions were considered in the case studied.

As the choice of a suitable function w preconditions the accuracy of the whole solution, the problem was at first dealt with—in the preparatory stage—by means of the linear theory of web buckling. A number of alternatives for a 6-term assumption for the buckled pattern, w, were considered.

The conclusions showed this optimum 6-term assumption:

$$w = f_1 \sin \frac{\pi x}{2a} \sin \frac{3\pi y}{b} + f_2 \sin \frac{3\pi x}{2a} \sin \frac{\pi y}{b}$$

$$+ f_3 \sin \frac{2\pi x}{a} \sin \frac{2\pi y}{b} + f_4 \sin \frac{\pi x}{2a} \sin \frac{\pi y}{b} + f_5 \sin \frac{\pi x}{a} \sin \frac{2\pi y}{b}$$

$$+ f_6 \sin \frac{3\pi x}{2a} \sin \frac{3\pi y}{b} \tag{4.8}$$

where f_i are unknown parameters.

The boundary value problem (4.1), (4.2)–(4.7) was then solved by means of the Papkovitch approach, i.e. the compatibility equation (eqn (4.1a)) was solved analytically whereas the equilibrium equation (eqn (4.1b)) was dealt with approximately via the energy method. A detailed description of the solution can be found in Massonnet et al. (1968), Škaloud et al. (1967) and Škaloud (1970); here only its main features are presented.

The shear load was measured by the ratio τ/τ_{cr}^*, where τ_{cr}^* is the critical stress of the corresponding unstiffened web.

On the question of the stiffening of the web, it was assumed that the stiffener was double-sided and symmetric with respect to the web. Its flexural rigidity was measured by the ratio γ/γ^*.

For the case under consideration (see Table 4.1)

$$\gamma^* = \frac{5 \cdot 4}{\beta} \left(\frac{2}{\beta} + \frac{2 \cdot 5}{\beta^2} - \frac{1}{\beta^3} - 1 \right)$$

where β is the aspect ratio of the whole web. For a square web, $\beta = 1$ and $\gamma^* = 13 \cdot 5$.

The determination of the deflection surface of the stiffened web in the post-buckled range involves evaluating the parameters $f_1, \ldots, f_6$. They follow from the solution to the system of 6 cubic equations

$$\frac{\partial \mathscr{E}}{\partial f_i} = 0 \qquad i = 1, \ldots, 6 \tag{4.9}$$

where $\mathcal{E}$ is the potential energy stored in the stiffened web, which can be written as a sum of the strain energy V (consisting of the energy V_m due to membrane stresses, of the strain energy V_b of bending) and T of the external forces. A detailed description of the involved set of formulae is given by Massonnet *et al.* (1968) and Škaloud *et al.* (1967).

System (4.9) was solved, for various τ/τ_{cr} and γ/γ^*, using a method of successive approximations similar to Newton's method of tangents. The parameters f_i being known, the deflection surface of the web (given by eqn (4.8)) can be found out. Thus the values of w were calculated at points having the following coordinates:

$$\frac{x}{a} = 0,\ 0.25,\ 0.5,\ 0.75,\ 1,\ 1.25,\ 1.5,\ 1.75,\ 2$$

$$\frac{y}{a} = 0,\ 0.125,\ 0.25,\ 0.375,\ 0.5,\ 0.625,\ 0.75,\ 0.875,\ 1$$

and for rigidities

$$\frac{\gamma}{\gamma^*} = 0,\ 0.125,\ 0.25,\ 0.5,\ 1,\ 1.5,\ 2,\ 3$$

The relative deflection, v_{c_s}/t, of the stiffener at the point C_s (Fig. 4.6):

$$\frac{v_{c_s}}{t} = -f_1' - f_2' + f_4' + f_6' \qquad (4.10)$$

where $f_i' = f_i/t$, is plotted for various γ in Fig. 4.7. This figure shows the deformation of the stiffener in the post-critical range.

The contour maps of the buckled pattern of the whole web, corresponding to

(a) $\gamma/\gamma^* = 0$ and $\tau/\tau_{cr}^* = 2$
(b) $\gamma/\gamma^* = 1$ and $\tau/\tau_{cr}^* = 3.8$
(c) $\gamma/\gamma^* = 3$ and $\tau/\tau_{cr}^* = 5.9$

are plotted in Fig. 4.8. They demonstrate the role of the flexural rigidity of the stiffener in the post-buckled range.

It follows from Fig. 4.9 that there is an analogy between the behaviour of a stiffener in the post-critical range and the large deflections of a compressed column. Therefore, it can be assumed, despite the fact that the

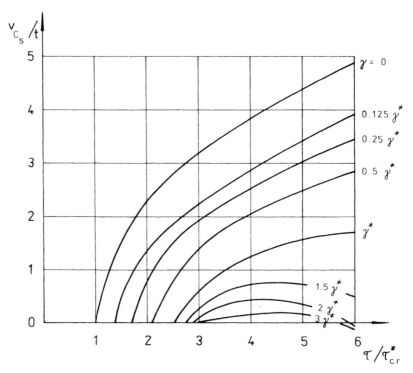

FIG. 4.7. Deflection of the transverse stiffener in terms of: (a) stiffener rigidity γ;
(b) load factor τ/τ_{cr}.

behaviour of an initially imperfect web was not studied numerically, that
the deflection of an initially curved stiffener is represented by the curves
given in Fig. 4.9(b), which are analogous to the curves shown in Fig. 4.9(a)
and valid for a compressed column. From the foregoing reasoning it can be
concluded that for a web with initial 'dishing' the stiffener (and, of course,
also the web) would deflect immediately from the beginning of loading.

An analysis of Fig. 4.8(a)–(c) shows that the efficacy of a vertical rib of
flexural rigidity equal to γ^* is limited in the sense that such a stiffener
deflects in the post-critical range with the buckling web sheet; hence, it
cannot be assumed that such a stiffener provides the web with rigid support.

If it is required that the stiffener should remain practically straight in the
whole post-buckled range, it is necessary to increase its flexural rigidity in
accordance with the following formula:

$$\gamma_0 = 3\gamma^* \tag{4.11}$$

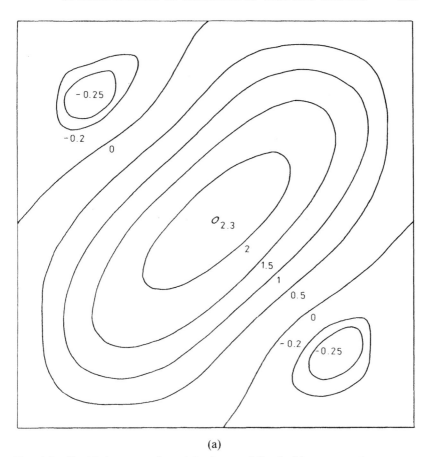

(a)

FIG. 4.8.   Buckled pattern of a web in shear and fitted with one central transverse stiffener: (a) $\gamma/\gamma^* = 0$, $\tau/\tau_{cr} = 2$.

4.4.3 Experimental Investigation into the Post-Buckled Behaviour of Stiffened Webs

In addition to the above theoretical results, some reliable experimental evidence was needed to validate the theory about the optimum rigidity of stiffeners on ordinary steel-plated structures. A wide range of valuable data was available from the tests by Massonnet and his associates (Massonnet, 1954; Massonnet *et al.*, 1962). However, other important gaps needed investigations.

For this reason R. Owen, K. C. Rockey and the author of this chapter

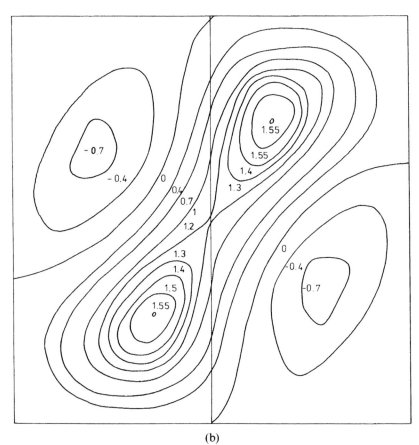

(b)

FIG. 4.8—*contd.*   (b) $\gamma/\gamma^* = 1$, $\tau/\tau_{cr} = 3\cdot8$.

decided to carry out another series of tests on plate girder webs fitted with stiffeners. The objectives of the experiment were as follows:

i.   To study the behaviour under pure bending of large web plates reinforced by one line of single-sided stiffeners placed at the optimum position.

ii. To study the behaviour of web plates reinforced by two lines of single-sided longitudinal stiffeners again placed at the optimum positions.

Since the overall objective of the investigation was to study the ultimate load behaviour of longitudinally reinforced girders when subjected to pure

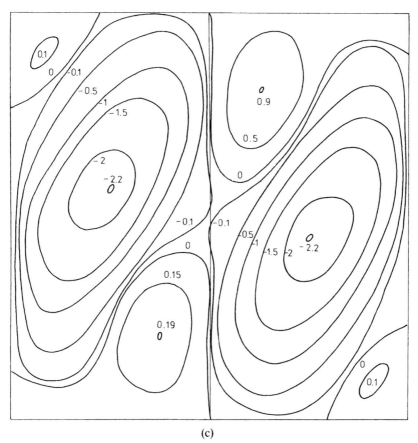

(c)

FIG. 4.8—*contd.*   (c) $\gamma/\gamma^* = 3$, $\tau/\tau_{cr} = 5\cdot9$.

bending, only that portion of the test girders so loaded was of interest. It was therefore decided to test girders having detachable end panels. These end panels were overdesigned to ensure that failure took place in one of the three panels, which were subjected to uniformly applied bending moment.

The general details of the girders tested are given in Fig. 4.10. Three types of girders were tested. One girder, TG 0, was fitted with only transverse stiffeners which had been designed to provide a flexural rigidity of $3\gamma^*$. The purpose of this was two-fold: first to provide a datum against which to gauge the performance of the longitudinally stiffened girders and secondly to check whether or not the 'efficiency factor' of 3 for transverse stiffeners,

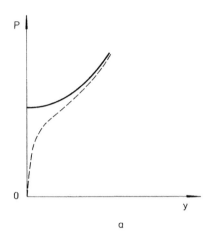

a

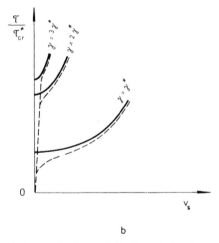

b

Fig. 4.9.   Analogy between the large deflections behaviour of: (a) a compressed
bar; (b) a transverse stiffener in a web in shear.

resulting from the above theoretical study (see Section 4.4.2), is
satisfactory.

The second series comprised four girders having the central test section
reinforced by one line of single-sided flat longitudinal stiffeners together
with two transverse stiffeners as shown in Fig. 4.10. The size of the
transverse stiffeners was kept constant and the girders only differed from
each other in the strength of the longitudinal stiffeners. Three of the girders

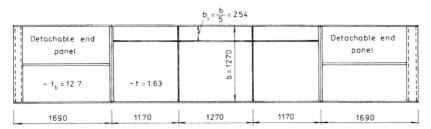

FIG. 4.10.   Experimental girders tested by Owen, Rockey and the author.

were fitted with single, one-sided longitudinal stiffeners and the fourth with a single, double-sided stiffener. The purpose of testing the latter girder was to provide a datum against which to judge the performance of the one-sided stiffeners. The actual (i.e. based on measured dimensions) values of the ratio γ/γ^* for the ribs used are in Table 4.2.

These longitudinal stiffeners were positioned at $\frac{1}{5}$ of the overall depth, b, of the girder from the compression flange, which is the optimum position—in accordance with the linear buckling theory (Dubas, 1954)—for a web which is simply supported on all four edges.

The third test series consisted of five girders, each having two lines of one-sided flat longitudinal stiffeners, whose actual γ/γ^*-ratios are also given in Table 4.2. The longitudinal stiffeners were positioned so as to give the maximum resistance against buckling. Thus the stiffeners were placed at $0 \cdot 123b$ and $0 \cdot 275b$ from the compression flange, following the conclusion of the study (Rockey and Cook, 1965).

During each test the deflections of the flanges, stiffeners and web sheet were carefully measured. Each girder was also instrumented with electrical

TABLE 4.2

Test girder	γ/γ^*
TG 0	no longitudinal stiffener
TG 1-1	0·89
TG 2-1	1·63
TG 3-1	5·80
TG 4-1	3·28
TG 1-2	0·64
TG 2-2	1·51
TG 3-2	2·94
TG 4-2	5·11
TG 5-2	7·75

resistance strain gauges, which made it possible to evaluate the stress state in the test girder.

In this chapter only the main test results are presented. For other experimental data, and a detailed description of the experimental set-up and apparatus, the reader is referred to Škaloud (1970).

The most important test conclusions are as follows:

The first test was conducted on girder TG 0, which did not have any longitudinal stiffener.

The measurements of the lateral deflection of the two transverse stiffeners in the central section indicated that only slight bowing of them occurred during loading; this demonstrated that the flexural rigidity of $3\gamma^*$ was sufficient to provide the web sheet with practically rigid support in the whole post-buckled range and, consequently, complied with the requirement of the optimum rigidity.

The girder failed at a load of 482·5 kN when the compression flange buckled vertically inwards. This load was 25·4 × the web-buckling load calculated by linear buckling theory on the assumption of simply-supported edges.

The effect of the size of the longitudinal ribs was demonstrated on girders TG 1-1, TG 2-1 and TG 3-1.

Girder TG 1-1 was fitted with one longitudinal stiffener of relative flexural rigidity $\gamma = 0.89\gamma^*$. The stiffener had an initial deformation, single curvature in form, with a maximum deflection 2·89 × the web thickness. Figure 4.11(a) gives the additional lateral deflections of the stiffener which occurred at different values of the applied load.

Failure occurred when the applied load had reached 555 kN, the compression flange again buckling inwards. The buckle in the flange formed above the region where the web deflections were largest.

Girder TG 2-1 had a single longitudinal stiffener whose relative flexural rigidity $\gamma = 1.63\gamma^*$ and, therefore, 1·84 × γ of the stiffener on girder TG 1-1. This girder behaved in a manner similar to that of girder TG 1-1, but the stiffener deflections (see Fig. 4.11(b)) were considerably smaller (less than $\frac{1}{2}$) than those of the rib on girder TG 1-1.

Girder TG 2-1 finally failed in the central panel by inward buckling of the compression flange at a load of 577·5 kN, the location of the flange buckle again coinciding with a large web buckle as in the case of girder TG 1-1. It is seen that by increasing the size of the longitudinal stiffener the failure load of the girder was enlarged by some 4%.

Girder TG 3-1, which was reinforced by an even heavier longitudinal rib ($\gamma = 5.8\gamma^*$), behaved similarly to the other two girders having a single

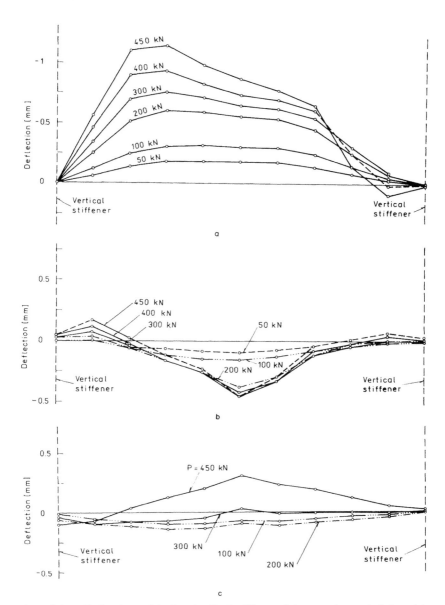

FIG. 4.11.   Deflections of the longitudinal stiffener of the experimental girders: (a) TG 1-1; (b) TG 2-1; (c) TG 3-1.

longitudinal stiffener, but the deflections of the stiffener were even smaller. Figure 4.11(c) gives the additional lateral deflections of the stiffener at various applied loads. It will be noted there that these deflections are extremely small, indicating that the stiffener was remaining practically straight. The strain gauge readings also indicated that very little bending of the stiffeners occurred. The girder failed at an applied load of 597·5 kN. This load is 24 % greater than the collapse load of girder TG 0.

The behaviour of the girders fitted with two lines of single-sided stiffeners was similar in character to that of girders with a single longitudinal stiffener. All girders failed by inward buckling of the compression flange. It was also noted that with an increase in the size of the longitudinal rib there was a decrease in the stiffener deflection and an increase in the ultimate load of the test girder. For girder TG 5-2, with stiffeners having $\gamma/\gamma^* = 7·75$, the deflection of the longitudinal stiffeners was so small that it could be assumed that they provided the web sheet with rigid support.

The conclusions can now be summed up as follows:

a. For a web subject to in-plane bending and reinforced by one single-sided open-section longitudinal stiffener, located at $\frac{1}{5}$ of the web depth, b, a relative flexural rigidity, $\gamma \simeq 6\gamma^*$ was sufficient for the stiffener to behave as a practically rigid one.

b. For a web subject to in-plane bending and reinforced by two equal single-sided open-section longitudinal stiffeners, situated at $0·123b$ and $0·275b$ from the compression flange, a relative flexural rigidity $\gamma \simeq 8\gamma^*$ sufficed to make the stiffeners behave as practically rigid ones.

Does this mean that the optimum stiffener rigidities should be defined so as to equal the above values?

Figure 4.12 gives the relationship between the percentage gain in the ultimate load of the test girders over that of girder TG 0 and the rigidity ratio of their longitudinal ribs. Also shown plotted is the corresponding percentage increase in weight due to a larger size of the longitudinal stiffeners.

The gain in ultimate strength for girders fitted with rigid longitudinal ribs is seen to be quite considerable. For example, in the case of girder TG 5-2 it amounts to 34·5 %. Associated with this 34·5 % increase in ultimate load is only a 5 % increase in weight.

Further, an examination of Fig. 4.12 shows that for larger values of γ/γ^* the character of the relationship between the gain in ultimate strength and the ratio γ/γ^* is rather flat. This indicates that, in this range, in spite of a

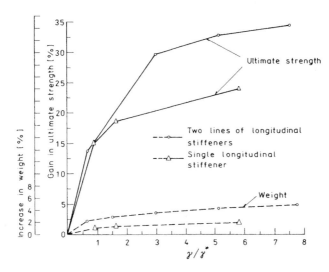

FIG. 4.12.   Gain in ultimate strength (and corresponding percentage increase in weight) in terms of stiffener rigidity.

very significant enlargement of stiffener rigidity only a slight gain in load-carrying capacity can be obtained; in other words, a considerable reduction of the stiffener parameter γ leads merely to a very small drop in the ultimate load of the girder.

Thus a lower value of γ than that given above for practically straight stiffeners can be considered in a definition of the optimum rigidity of open-section longitudinal stiffeners. An analysis of Fig. 4.12 shows that a value

$$\gamma_0 = 4\gamma^* \qquad (4.12)$$

can satisfy this purpose.

4.5   RECOMMENDATION FOR PRACTICAL DESIGN

Maquoi *et al.* (1981) summarised the available theoretical and experimental evidence in regard to the behaviour of stiffened webs and flanges, and formulated this recommendation for the design of rigid stiffeners:

When the stiffeners are to be designed so as to provide the web or flange sheet with rigid support, their optimum rigidity, γ_0, is given by

$$\gamma_0 = k\gamma^* \qquad (4.13)$$

where γ^* can be taken from Table 4.1 for the corresponding type of stiffening and loading and the factor k, reflecting the effect of initial imperfections and of post-buckling behaviour, is as follows:
For transverse stiffeners:

$$k_t = 1 \quad \text{for} \quad b/t \le 75 \tag{4.14a}$$

$$k_t = 3 \quad \text{for} \quad b/t \ge 150 \tag{4.14b}$$

with linear interpolation for $75 < b/t < 150$; b/t is the depth-to-thickness ratio of the whole web.
For longitudinal stiffeners:
  (i) of open cross-sections (flats, angles, T-shapes, etc.)

$$k_l = 1 \cdot 25 \quad \text{for} \quad b/t \le 120 \tag{4.15a}$$

$$k_l = 4 \quad \text{for} \quad b/t \ge 240 \tag{4.15b}$$

with linear interpolation for $120 < b/t < 240$.
  (ii) of closed cross-sections

$$k_l = 1 \cdot 25 \quad \text{for} \quad b/t \le 120 \tag{4.16a}$$

$$k_l = 2 \cdot 5 \quad \text{for} \quad b/t \ge 240 \tag{4.16b}$$

with linear interpolation for $120 < b/t < 240$.
When the web is stiffened by more than one longitudinal stiffener, the above values of k_l can be multiplied by a reduction factor $m = 0 \cdot 8$.
An inspection of eqns (4.14)–(4.16) shows that they correctly reflect:

a. the theoretical and experimental conclusions described above in Section 4.4;
b. the fact that post-buckled behaviour (and, consequently, the corresponding demand on stiffener rigidity) grows with the web depth-to-thickness ratio;
c. the evidence that closed-section ribs are considerably more efficient than open-section ones, requiring an optimum rigidity equal only to about 60% of that associated with open-section stiffeners.

When the web is subjected to a combination of compression σ_c, bending σ_b and shear τ, the value γ_l^* for the longitudinal stiffeners is calculated by means of the following interaction formula:

$$\gamma_l^* = \sqrt{\left(\left[\gamma_{l,\sigma}^*\left(\frac{\sigma_c}{\sigma_{c,ult}} + \frac{\sigma_b}{\sigma_{b,ult}}\right)\right]^2 + \left(\gamma_{l,\tau}^*\frac{\tau}{\tau_{ult}}\right)^2\right)} \tag{4.17a}$$

Here the quantity $\gamma_{l,\sigma}^*$ is related to the longitudinal normal stress $\sigma = \sigma_c + \sigma_b$ and $\gamma_{l,\tau}^*$ to the shear stress τ. The quantities $\sigma_{c,ult}$, $\sigma_{b,ult}$ and τ_{ult} are ultimate loads of the web panel under consideration if loaded in pure compression, bending or shear, respectively.

If the quantity $\gamma_{l,\sigma}^*$ is not known, but we know the quantities γ_{l,σ_c}^* and γ_{l,σ_b}^* for pure compression and bending, respectively, eqn (4.17a) can be replaced by the following one:

$$\gamma_l^* = \sqrt{\left[\left(\gamma_{l,\sigma_c}^* \frac{\sigma_c}{\sigma_{c,ult}} + \gamma_{l,\sigma_b}^* \frac{\sigma_b}{\sigma_{b,ult}}\right)^2 + \left(\gamma_{l,\tau}^* \frac{\tau}{\tau_{ult}}\right)^2\right]} \qquad (4.17b)$$

Equation (4.17a) or eqn (4.17b) is applied to each of the two adjacent panels of the longitudinal stiffener under consideration, and the larger value of γ_l^* is then inserted into eqn (4.13) and considered in design.

In the design of transverse stiffeners, only shear loading, τ, is taken into account in the calculation; thus the value of $\gamma_{t,\tau}$ can be found straight away in Table 4.1; usually the formula for an infinite plate with equally spaced identical stiffeners is applied, since it best approximates the frequently encountered case of a plate girder fitted with numerous equidistant vertical ribs. The presence of longitudinal stiffeners is then accounted for by conferring a fictitious thickness, t_f, on the infinite plate so that the ultimate shear load of the substituted plate is equal to the ultimate shear load of the original stiffened plate.

Then, for a longitudinally stiffened web,

$$\gamma^* = \gamma_{t,0}^* \left(\frac{t_f}{t}\right)^3 \qquad (4.18)$$

where $\gamma_{t,0}^*$ is found for a similar web without longitudinal ribs.

4.6 OTHER ASPECTS OF STIFFENER DESIGN

The design approach above is based merely on rigidity considerations. But, design of stiffeners from the point of view of strength is also important and there are countries where the design of stiffeners entirely follows this view point.

The strength design requires that the dimensions of stiffeners be such that the stiffeners can sustain all forces which act on them as a result of sheet buckling or are transferred into them from the external loading. This analysis is of particular importance in the design of the transverse stiffeners and end posts of steel-plate girders.

However, this aspect of stiffener behaviour is beyond the scope of this chapter, which is only concerned with the question of optimum rigidity.

4.7 CONCLUSIONS

The idealised optimum rigidity γ^*, resulting from the linear theory of plate buckling, can be defined in three different ways. A set of formulae for γ^* for the most frequently encountered cases of stiffening and loading is listed in Table 4.1.

When this quantity γ^* is checked in the light of theoretical and experimental evidence regarding the post-buckled behaviour of stiffened webs, it is found that the rigidity γ^* does not suffice for the stiffeners to remain straight in the post-critical range and thus provide the buckling sheet with rigid support.

In order to achieve this objective, the flexural rigidity of stiffeners has to be enlarged to a value $\gamma_0 = k\gamma^*$, where the factor k (> 1) expresses the effect of initial imperfections and post-buckled behaviour.

The factor k is a function of:

the kind of stiffeners (transverse or longitudinal ones);
the type of the cross-section of stiffeners (open or closed ones);
the depth-to-thickness ratio of the web.

A proposal for the k-values to be employed in practical design is given in Section 4.5.

REFERENCES

DUBAS, CH. (1954) Le voilement de l'âme des poutres fléchies et raidies au cinquième supérieur. *Mémoires de l'AIPC*, **14**, 1–12.

KLÖPPEL, K. and MÖLLER, K. H. (1968) *Beulwerte ausgesteifter Rechteckplatten, Band 2*, W. Ernst u. Sohn, Berlin.

KLÖPPEL, K. and SCHEER, J. (1960) *Beulwerte ausgesteifter Rechteckplatten, Band 1*, W. Ernst u. Sohn, Berlin.

KŘÍSTEK, V. and ŠKALOUD, M. (1977) Solution to the stability problem of longitudinally stiffened compression flanges by folded plate theory. *Acta technica ČSAV*, No. 4, 387–412.

MAQUOI, R., MASSONNET, CH. and ŠKALOUD, M. (1981) Design of stiffened webs. *Stavebnicky časopis (Building Journal)*, No. 2, 73–88.

MASSONNET, CH. (1954) Essai de voilement sur poutres à âme raidie. *Mémoires de l'AIPC*, **14**, 125–86.

MASSONNET, CH., MAS, E. and MAUS, H. (1962) Essai de voilement sur deux poutres à membrures et raidisseurs tubulaires. *Mémoires de l'AIPC*, **22**.

MASSONNET, CH., ŠKALOUD, M. and DONÉA, J. (1968) Comportement postcritique d'une plaque carrée raidie cisaillée uniformément. Deuxième partie: Repartition des contraintes et analyse de l'état limite. *Mémoires de l'AIPC*, **28**, 137–56.

ROCKEY, K. C. and COOK, I. T. (1965) The buckling under pure bending of a plate girder reinforced by multiple longitudinal stiffeners. *Int. J. of Solids and Structures*, **I**, 79–92, 147–56.

ŠKALOUD, M. (1970) Post-buckled behaviour of stiffened webs. *Trans. Czech. Acad. Sci., Techn. Sci. Ser.*, **80**(1), 1–154.

ŠKALOUD, M. and KŘÍSTEK, V. (1977). Folded plate theory analysis of the effect of the stiffener configuration upon the buckling of longitudinally stiffened compression flanges. *Acta technica ČSAV*, No. 5, 577–601.

ŠKALOUD, M. and KŘÍSTEK, V. (1981) Stability problems of steel box-girder bridges. *Trans. Czech. Acad. Sci., Techn. Sci. Ser.*, **91**(1), 1–122.

ŠKALOUD, M., MASSONNET, CH. and DONÉA, J. (1967) Comportement post-critique d'une plaque carrée raidie cisaillée uniformément. Première partie: Solution générale et déformée de la plaque. *Mémoires de l'AIPC*, **27**, 187–210.

ŠKALOUD, M. and ZÖRNEROVÁ, M. (1977) Linear buckling theory optimum rigidity of the longitudinal stiffeners of the compression flanges of steel box-girder bridges. *Acta technica ČSAV*, No. 1, 33–51.

Chapter 5

ULTIMATE CAPACITY OF STIFFENED PLATES IN COMPRESSION

N. W. Murray

Department of Civil Engineering, Monash University, Victoria, Australia

SUMMARY

Some of the important factors which influence the collapse load of a stiffened plate panel in compression are the conditions of support of the edges, the effects of local and global elastic buckling, the effects of initial welding stresses and initial geometrical imperfections, the magnitude of the yield stress and the shape and size of the stiffeners. A comprehensive method which considers all of these parameters does not yet exist. This chapter describes some simplified methods for predicting the collapse load of stiffened plate panels. Some of these model the panel as a column but others treat it as an orthotropic plate.

NOTATION

A	= area of cross-section
A_1	= maximum height of a buckle
b	= width of isolated plate or a plate element in the cross-section
b_e	= effective width of plate after buckling
B	= width of stiffened plate panel
C	= parameter defined in eqn (5.32)
D_x, D_z	= section properties defined in eqn (5.11)
E, E_x, E_z	= Young's modulus for isotropic and anisotropic plates

G, G_{xz}	= shear modulus of isotropic and anisotropic plates
H	= section property defined in eqn (5.11)
I	= second moment of area of the cross-section
k	= dimensionless factor in buckling formula (eqn (5.24))
K_{bs}	= secant effective width factor (eqn (5.29))
K_{xz}	= shearing rigidity of an equivalent plate
L	= length of panel
L_{cr}	= length of buckle
m	= integer (eqns (5.14) and (5.15))
m_1	= magnification factor (eqn (5.31))
M'_p	= plastic moment per unit width of plate with hinge perpendicular
M''_p	= plastic moment per unit width of plate with hinge inclined to load direction
M_x, M_z, M_{xz}	= bending and twisting moments per unit length
n	= integer (eqns (5.14) and (5.15), or constant (eqn (5.25))
N_x, N_z, N_{xz}	= tensile in-plane forces per unit length
P_z	= axial force per unit length acting in the z direction
r	= radius of gyration of the cross-section $= (I/A)^{1/2}$
t	= thickness of plate
w	= out-of-plane surface loading per unit area
w_0	= out-of-plane deflection at centre of a plate measured from initial shape
W	= force acting in direction of the normal (Fig. 5.2)
y	= initial out-of-plane deflection of a plate
y_0	= initial out-of-plane deflection at centre of a plate
Y	= out-of-plane surface loading per unit area
Y_0	= uniform surface loading
y_1, y_2	= distances from neutral axis to mid-plane of plate and tip of stiffener, respectively (Fig. 5.2)
α	= aspect ratio of plate $= L/b$
β	= angle between plastic hinge and line perpendicular to load direction
δ_c	= total central deflection of panel after axial stress is applied
δ_0	= initial central deflection of panel
δ_{c0}	= total central deflection of panel before axial stress is applied
ε	= axial strain
η	= imperfection factor in Perry–Robertson formula

v, v_x, v_z	= Poisson's ratio for isotropic and anisotropic plates
σ	= axial stress
σ_{av}	= average stress in plate after buckling
σ_{cr}	= critical value of σ
σ_e	= stress along edge of plate after buckling (Fig. 5.5)
σ_E	= Euler buckling stress = $\pi^2 E (L/r)^{-2}$
σ_m	= average axial stress at the point of failure
$\sigma_{m\ av}$	= average value of σ_{pm} for whole cross-section of panel
σ_{pm}	= average axial stress of an individual plate element at its point of failure
σ_y	= yield stress
Φ	= stress function

5.1   INTRODUCTION

A thin steel plate is very flexible when it carries loads which act in the direction of its normal, but it is extremely stiff when the loads are applied within its plane. It is this rigidity which engineers attempt to utilise when they design thin-walled structures such as plate girders, box girders, box columns and so on. With these structures designers have to restrain the plates, which form the structures, against out-of-plane distortion or buckling because of their flexibility in that direction. This restraint can be provided by folding the plate along parallel lines which lie in the longitudinal direction. Such structures are called folded plates if they are wide or, if they are narrow and rolled from a flat strip, cold-formed sections, but these structures are outside the scope of this chapter. An alternative way of restraining a plate is to provide longitudinal stiffeners, which are additional plate elements whose planes are inclined to that of the plate. Thus the concept of a stiffened plate has been developed and now a stiffened plate panel forms the basic building block of many thin-walled steel structures. The individual flat plates which form the cross-section of the panel are referred to here as elements. There are many forms of stiffeners and Fig. 5.1 illustrates some of the more common types used in the construction industry. Those shown in Fig. 5.1(a)–(d) are generally used for applications in which the applied loading is in the plane of the panel such as in box columns and in the webs and bottom flanges of box girders, while those shown in Fig. 5.1(e)–(g) are generally used where both axial and normal loading are important. An example of the latter application is in the upper deck of a box-girder bridge where the panel must

N. W. MURRAY

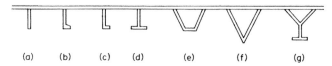

FIG. 5.1.   Various shapes of stiffeners which have been used in stiffened plate construction.

simultaneously carry axial compression due to overall bending of the girder and normal loading due to the wheel loads of the passing vehicles. These stiffened panels with multiply-connected or so-called closed cross-sections have considerably greater resistance to twisting moments than those with single-connected or so-called open sections. Thus the local distortions of closed cross-sections are less and this has advantages for bridge decks because there is less likelihood of cracking of the pavement under wheel loads.

The purpose of this chapter is to show ways of determining the failure loads of stiffened plate panels but in order to do this it is first necessary to understand how they behave as the loads increase from zero up to the point of collapse. Panels behave and collapse in a variety of ways depending upon their geometrical and loading configurations, the manner in which they are supported, the nature and magnitude of their initial imperfections, the type of steel used and so on. To try to understand panel behaviour we shall start with the simplest case first, viz., panels with stocky cross-sections, and later deal with panels whose plate elements are wider and thinner. Thus, in Section 5.2 panels which can be analysed as wide beams or columns are treated. With a small increase in complexity a panel with a stocky cross-section can be analysed as if it were an orthotropic plate (Section 5.3). These analyses enable approximate values of various internal actions (shear forces, bending moments, etc.) to be obtained and they can then be used with an appropriate criterion of failure to estimate the failure load. These methods of analysis have a number of limitations especially when the loading is predominantly in-plane. In such cases these theories, which treat a panel as an equivalent orthotropic plate, cannot deal with local buckling of the individual elements of the cross-section. Local buckling of a panel entails localised and often severe distortion of the cross-section whereas with global buckling, although there may be some small distortion of the cross-section, the out-of-plane distortions over large adjacent areas of the panel are in the same direction. Orthotropic plate theory can give good estimates of global buckling stresses but it ignores local buckling stresses. When a panel has a cross-section with plate elements, which are supported in some way around each edge and with individual b/t ratios of, say, 45 or more, local buckling may occur at a lower stress than the yield stress. The strategy for determining the collapse load, therefore, has to change. For panels with these higher b/t ratios it is necessary first to evaluate the critical stress σ_{cr} at which local buckling occurs and then it is possible to estimate the failure stress by using various empirical and semi-empirical methods (Section 5.4).

Another important problem is to evaluate the failure load of a panel in the upper deck of a box-girder bridge. As stated earlier, it must carry simultaneously axial loading and transverse wheel loading. If the panel has a stocky cross-section the methods used in Sections 5.2 and 5.3 can be used, but for panels susceptible to local buckling, other criteria have to be used. Section 5.5 briefly describes investigations of this class of problem.

Even with the most careful control on fabrication procedures it is not possible to manufacture a panel without some initial distortion and initial stresses. Because of the influence of these parameters, two nominally identical panels can behave in quite different ways. As a simple example, a wide panel with axial loading can behave like an Euler column. If it is initially bowed towards the plate, the stiffeners will carry higher compressive stresses than the plate and the failure mode will be one which involves a crumpling or plastic mechanism of the stiffeners. On the other hand, if the initial bow is towards the stiffeners, the plate will probably develop a plastic mechanism first and the failure load will be different. Not only will this be so, but perhaps what is more important, collapse will normally be much more sudden in the former example and the designer should apply higher factors of safety. From a designer's viewpoint post-buckling behaviour of a panel is important because some panels collapse in a very 'brittle' manner without warning while for others failure is very 'ductile'. This does not mean that the steel used is brittle or ductile; it is simply a function of the geometric parameters. Ductile panels not only give warning of local distress but also allow redistribution of the internal forces to other regions which are less highly loaded. Ways of examining post-buckling behaviour are described in Section 5.6.

The foregoing description of stiffened plate behaviour has been greatly simplified in order to highlight certain aspects of this fascinating field of study. Initial imperfections, welding stresses and other factors tend to introduce a certain amount of uncertainty about the behaviour of most panels. During carefully conducted laboratory tests one observes many phenomena which are at first surprising and difficult to understand. A panel under axial loading may start to deflect towards the stiffeners and then suddenly it will fail in the opposite direction. Investigation shows that one of the stiffeners had reached its point of failure even though the initial bowing was in the sense which would relieve the stiffener to some extent. Failure of one stiffener can sometimes cause a chain-reaction failure in other stiffeners if their behaviour is brittle. Although research is at the stage where the theoretical behaviour of a given panel will follow fairly closely that observed by an experiment, it is not usually feasible to use these theoretical techniques for design purposes. In the following sections it will be seen that empirical rules have to be used to a greater or lesser extent in each of the analytical methods described. In the final section (Section 5.7) the ways that the theoretical results and empirical rules have been brought together to form simplified design rules in a few representative design codes are described.

5.2   COLLAPSE OF PANELS WITH STOCKY CROSS-SECTIONS AND FREE EDGES

The simplest panel to analyse is that shown in Fig. 5.2(a) when the individual elements have low b/t ratios and the edges of the panel are free. Each inner plate element is restrained against out-of-plane deflections by stiffeners; this means that b/t can be as high as about 35. However, when an element has a free edge the b/t ratio will have to be much less (say 10) if it is to be considered stocky. Within these constraints the panel is unlikely to suffer from local buckling and the whole cross-section is effective in resisting bending. If the panel described carries normal and axial loads simultaneously it can be analysed as a wide column. The following example illustrates how this class of problem may be treated.

The wide and stocky panel with initial central deflection δ_0 and with pinned ends shown in Fig. 5.2(b) is loaded axially by a stress σ and a normal force W at mid-point. If W acted alone the central deflection would be

$$\delta_{c0} = \delta_0 + \frac{WL^3}{48EI} \qquad (5.1)$$

After σ is applied the central deflection is magnified by the factor $(1 - \sigma/\sigma_E)^{-1}$ where σ_E is the Euler buckling stress of the panel as a long column (i.e. $\pi^2 E(L/r)^{-2}$). Thus after the application of both W and σ the total central deflection is

$$\delta_c = \frac{\delta_{c0}}{1 - \sigma/\sigma_E} = \frac{\delta_0 + \dfrac{WL^3}{48EI}}{1 - \sigma/\sigma_E} \qquad (5.2)$$

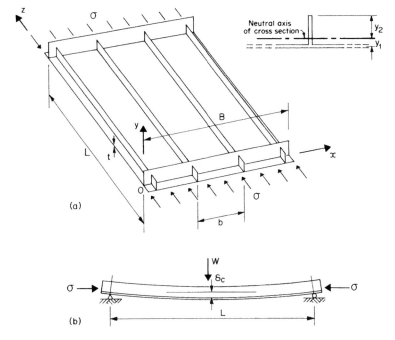

FIG. 5.2.   A wide stocky panel with free longitudinal edges behaves like a wide column.

The compressive stress in the middle plane of the plate is

$$\sigma_1 = \sigma - \frac{\sigma A \delta_c y_1}{I} \qquad (5.3)$$

and that at the tip of the stiffener is

$$\sigma_2 = \sigma + \frac{\sigma A \delta_c y_2}{I} \qquad (5.4)$$

One criterion of failure of such a panel is to define collapse as the point at which the maximum stress at the middle plane of the plate or at the outer tip of the stiffener just reaches the yield stress σ_y. At this point the average axial stress σ attains its maximum value denoted by the symbol σ_m. Thus, with σ_1 or σ_2 equal to σ_y and by using eqn (5.2), eqns (5.3) and (5.4) can be written as the following single equation

$$(\sigma_E - \sigma_m)(\sigma_y - \sigma_m) = \sigma_m \sigma_E \eta \qquad (5.5)$$

where

$$\eta = -\frac{A\delta_{c0}y_1}{I} \quad \text{or} \quad +\frac{A\delta_{c0}y_2}{I} \qquad (5.6)$$

depending upon whether the plate or stiffener, respectively, is being investigated. Equation (5.5) is a quadratic in the unknown σ_m and has the following solution

$$\sigma_m = \tfrac{1}{2}[\sigma_y + \sigma_E(1+\eta)] - \tfrac{1}{2}[\{\sigma_y + \sigma_E(1+\eta)\}^2 - 4\sigma_y\sigma_E]^{1/2} \qquad (5.7)$$

A load factor of say, 0·6 can be applied to this equation when it is used for the design of panels in this class. Equation (5.7) is almost identical to the well-known Perry–Robertson formula (Parkes, 1965) for the design of structural steel columns. The only difference is that in the latter formula it is assumed that $W = 0$ and η is related empirically to the slenderness ratio L/r of the column.

The above method for the design of stiffened plate panels has been verified experimentally to a limited extent (Michelutti, 1976; Michelutti and Murray, 1977). Experiments show that the method can also be applied to a panel with higher b/t ratios so long as the L/r ratio is sufficiently high because then local buckling does not occur or its effect is insignificant. The method has the important advantages that it is easy to apply and it uses an approach familiar to engineers. However, it should be used with caution because it ignores the influences of the initial deflection of the plate elements and residual welding stresses (see Sections 5.4 and 5.6) both of which reduce the stiffness of parts of the cross-section and therefore result in lower Euler buckling stresses.

Although thin-walled structures which do not buckle locally always have some reserves of strength above the load at which yielding commences, these margins are not usually as substantial and reliable as they are for normal structural steelwork. This is because once yielding commences in a thin-walled structure it often penetrates rapidly through the cross-section with increasing load until a local plastic mechanism is formed. Thus although the method appears at first sight to be very conservative this is not often the case.

5.3   ANALYSIS OF SIMPLY-SUPPORTED PANELS WITH STOCKY CROSS-SECTIONS AS ORTHOTROPIC PLATES

When the sides of a stocky panel are simply-supported the restraint there increases the failure load above that predicted by the theory of the previous section where it was assumed that the sides of the panel were unsupported.

It is of course assumed that the same failure criterion is used in both cases. A different approach is required in order to evaluate the maximum stress. In the method described in this section, the panel is treated as a simply-supported orthotropic plate. In such a plate the bending and in-plane stiffness properties are different in the two orthogonal directions x and z (Fig. 5.3). One of the most general analyses of such a plate was carried out by Rostovtzev (1940) who derived the following governing equations for a plate with initial out-of-plane imperfections $y(x, z)$ and which carries both in-plane forces (per unit length) N_z, N_x, N_{xz} and out-of-plane loading Y (per unit of area). As is usual in structural theory N_z and N_x are assumed to be positive if they are tensile.

$$D_x \frac{\partial^4 w}{\partial x^4} + 2H \frac{\partial^4 w}{\partial x^2 \partial z^2} + D_z \frac{\partial^4 w}{\partial z^4} = \frac{\partial^2 \Phi}{\partial z^2} \frac{\partial^2 (y+w)}{\partial x^2} + \frac{\partial^2 \Phi}{\partial x^2} \frac{\partial^2 (y+w)}{\partial z^2}$$

$$- 2 \frac{\partial^2 x}{\partial x \partial z} \frac{\partial^2 (y+w)}{\partial x \partial z} + Y \qquad (5.8)$$

$$\frac{1}{t_z E_z} \frac{\partial^4 \Phi}{\partial x^4} + 2 \left[\frac{1}{K_{xz}} - \frac{v_x}{t_x E_x} - \frac{v_z}{t_z E_z} \right] \frac{\partial^4 \Phi}{\partial x^2 \partial z^2} + \frac{1}{t_x E_x} \frac{\partial^4 \Phi}{\partial z^4}$$

$$= -\frac{\partial^2 y}{\partial z^2} \frac{\partial^2 w}{\partial x^2} + 2 \frac{\partial^2 y}{\partial x \partial z} \frac{\partial^2 w}{\partial x \partial z} - \frac{\partial^2 y}{\partial x^2} \frac{\partial^2 w}{\partial z^2} - \frac{\partial^2 w}{\partial z^2} \frac{\partial^2 w}{\partial x^2} + \left(\frac{\partial^2 w}{\partial x \partial z} \right)^2 \qquad (5.9)$$

where

Φ is a stress function with the following properties:

$$\frac{\partial^2 \Phi}{\partial x^2} = N_z \qquad \frac{\partial^2 \Phi}{\partial z^2} = N_x \qquad \text{and} \qquad \frac{\partial^2 \Phi}{\partial x \partial z^2} = - N_{xz}$$

$$(5.10)$$

t_x and t_z are the average thickness of the panel in the x and y directions, respectively;

K_{xz} is the shearing rigidity of the equivalent plate;

D_x $= (EI_x)/(1 - v_x v_z) = $ average flexural rigidity per unit width of the stiffened plate under bending moment M_x;

D_z $= (EI)_z/(1 - v_x v_z) = $ average flexural rigidity per unit width of the stiffened plate under bending moment M_z;

H $= \frac{1}{2}(v_x D_x + v_z D_z) + 2(GI)_{xz}$; (5.11)

$2(GI)_{xz} = (\partial^2 w/\partial x \partial z)/M_{xz} = $ average torsional rigidity per unit width $= G_{xz} t^3/12$;

v_x, v_z $= $ Poisson's ratio in the x and z directions, respectively;

E $= \sqrt{(E_x E_z)} = $ modified Young's modulus;

G_{xz} $= E/2(1 + \sqrt{(v_x v_z)}) = $ modified shear modulus.

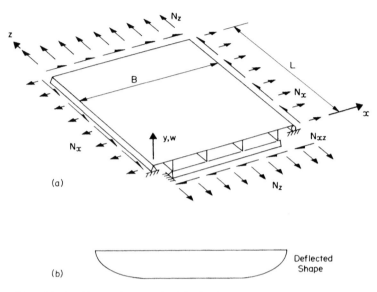

FIG. 5.3.   (a) A wide stocky panel with simply-supported sides behaves like an orthotropic plate. (b) Deflected shape of a stiffened panel suggests that it can be treated as a wide beam or column.

After solving eqns (5.8) and (5.9) for w the bending and twisting moments at any point are obtained from the following expressions:

$$M_x = -\frac{(EI)_x}{1 - v_x v_z}\left[\frac{\partial^2 w}{\partial x^2} + v_y \frac{\partial^2 w}{\partial z^2}\right]$$

$$M_z = -\frac{(EI)_z}{1 - v_x v_z}\left[\frac{\partial^2 w}{\partial z^2} + v_z \frac{\partial^2 w}{\partial x^2}\right] \tag{5.12}$$

$$M_{xz} = 2(GI)_{xz}\frac{\partial^2 w}{\partial x\,\partial z}$$

When the plate is isotropic its properties in the x- and z-directions are identical and eqns (5.8) and (5.12) simplify to the well-known Marguerre equations for an initially imperfect plate.

Although eqns (5.8)–(5.12) appear to be formidable they have been solved for a number of interesting cases by Basu and Chapman (1966) and by Aalami and Chapman (1969) by using the finite difference method. Such solutions require large computers and iterative methods because of the non-linearity of the equations. However, for special cases it is possible to obtain solutions by analytical methods, as is seen from the two following examples. The first deals with a panel with bending only and the second with a buckling problem.

Example 1: Initially perfect plate with uniform transverse loading only and simply-supported edges

In this case $y = \Phi = 0$ and eqn (5.9) is satisfied identically. Equation (5.8) becomes:

$$D_x \frac{\partial^4 w}{\partial x^4} + 2H \frac{\partial^4 w}{\partial x^2 \partial z^2} + D_z \frac{\partial^4 w}{\partial z^4} = Y \tag{5.13}$$

A uniform load Y_0 per unit of area of plate can be represented by a double sine series, viz.:

$$Y = \frac{16 Y_0}{\pi^2} \sum_{m=1,3,5}^{\infty} \sum_{n=1,3,5}^{\infty} \frac{1}{mn} \sin \frac{m\pi x}{B} \sin \frac{n\pi z}{L} \tag{5.14}$$

The equation, which results from combining eqns (5.13) and (5.14), is assumed to have a solution of the form

$$w = \sum_{m=1,3,5}^{\infty} \sum_{n=1,3,5}^{\infty} w_{mn} \frac{m\pi x}{B} \frac{n\pi z}{L} \tag{5.15}$$

After substitution w_{mn} is obtained directly, whence

$$w = \frac{16 Y_0}{\pi^6} \sum_{m=1,3,5}^{\infty} \sum_{n=1,3,5}^{\infty} \frac{\sin \dfrac{m\pi x}{B} \sin \dfrac{n\pi z}{L}}{mn \left[\dfrac{m^4}{B^4} D_x + \dfrac{2m^2 n^2}{B^2 L^2} H + \dfrac{n^4}{L^4} D_z \right]} \tag{5.16}$$

The deflection at $z = L/2$ of a typical stiffened plate with stiffeners in the z-direction only ($D_z \gg D_x$ and H) is shown in Fig. 5.3(b). It is seen that most of the plate deflects as a wide beam in a similar manner to an identical plate with free edge except in the vicinity of the sides. Numerical studies show that if the panel is wide, i.e. has five or more stiffeners, it can be treated as a wide beam and its behaviour is almost independent of the condition of support along each side.

Example 2: An infinitely long, initially perfect plate with simply-supported sides carries a compressive axial loading P_z per unit width of plate which acts in the z-direction

With $y = Y = 0$ and $\Phi = -P_z x^2/2$ eqn (5.9) simplifies to

$$\left(\frac{\partial^2 x}{\partial x\,\partial z}\right)^2 - \frac{\partial^2 w}{\partial x^2}\frac{\partial^2 w}{\partial z^2} = 0 \tag{5.17}$$

In the vicinity of the buckling load when the panel begins to deflect from its initially flat condition, all of the terms in eqn (5.17) are very small so the equation is satisfied identically. Equation (5.8) simplifies to

$$D_x \frac{\partial^4 w}{\partial x^4} + 2H \frac{\partial^4 w}{\partial x^2\,\partial z^2} + D_z \frac{\partial^2 w}{\partial z^2} = -P_z \frac{\partial^2 w}{\partial z^2} \tag{5.18}$$

This is an eigenvalue problem, i.e. it has the trivial solution $w = 0$ except when P_z assumes certain characteristic values. The lowest of these is the value of P_z at which the panel buckles. To find an approximation of this value of P_z, denoted by $P_{z\ cr}$, we assume a form of the buckling mode which satisfies the requirement that the boundaries are simply-supported as follows:

$$w = A_1 \sin\frac{\pi x}{B}\sin\frac{\pi z}{L} \tag{5.19}$$

where A_1 is the maximum height of the buckle and L is its length. Substitution into eqn (5.18) gives the following expression for $P_{z\ cr}$:

$$P_{z\ cr} = \frac{\pi^2}{B^2}\left[D_x\frac{L^2}{B^2} + 2H + D_z\frac{B^2}{L^2}\right] \tag{5.20}$$

It is interesting to notice for the case when the stiffening is mostly in the z-direction, i.e. D_x and H are small compared with D_z and if $L/B < 1$ that eqn (5.20) gives an expression for the buckling stress which is close to the Euler buckling stress. For many practical situations, especially when nearly all of the stiffening is in the z-direction, and if the panel is relatively wide ($L/B < 1$), the simple Euler formula can be used to estimate the critical stress and the value will be conservative. Furthermore, for these cases the failure load can be estimated by using the Perry–Robertson formula developed in the previous section.

When $L/B \gg 1$ this simplification is not possible because the term $D_x L^2/B^2$ in eqn (5.20) is then significant. Let us consider an infinitely long plate which is simply-supported along each side and let L be the length of

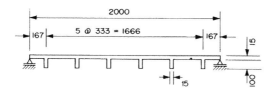

FIG. 5.4. Cross-section at a stiffened plate which is analysed as an orthotropic plate.

the buckles. It is required to find the value of L which minimises $P_{z\,cr}$. By differentiation of eqn (5.20) and equating to zero

$$\frac{L_{cr}}{B} = \left[\frac{D_z}{D_x}\right]^{1/4}$$ (5.21)

and a substitution into eqn (5.20) gives

$$P_{z\,cr} = \frac{2\pi^2}{B^2}\,[(D_xD_z)^{1/2} + H]$$ (5.22)

As a numerical example consider the cross-section shown in Fig. 5.4. It is required to evaluate its buckling load and the length of each buckle. ($E = 206\,000$ MPa, $v = 0.3$, $I_z = 30.727 \times 10^6$ mm^4). From eqns (5.11):

$$D_z = \frac{206\,000 \times 30.727 \times 10^6}{2000(1 - 0.3^2)} = 3477.93 \times 10^6 \text{ Nmm}$$

$$D_x = \frac{206\,000 \times 15^3}{12(1 - 0.3^2)} = 63.67 \times 10^6 \text{ Nmm}$$

$$H = \frac{0.3}{2}\,(3477.93 \times 63.67) \times 10^6 + \frac{80\,000 \times 15^3}{12} = 533.74 \times 10^6 \text{ Nmm}$$

From eqn (5.22):

$$P_{z\,cr} = \frac{2\pi^2}{2000^2}\,[(3477.9 \times 63.67 \times 10^{12})^{1/2} + 553.74 \times 10^6] = 5054.7 \text{ Nmm}^{-1}$$

Therefore

$$\text{critical load} = 2000 \times 5054.7 = 10.11 \times 10^6 \text{ N}$$

From eqn (5.21) the length of the buckles is

$$L_{cr} = 2000[3477.93/63.67]^{1/4} = 5437 \text{ mm}$$

An Euler column of this cross-section and length has a buckling load of $2 \cdot 11 \times 10^6$ N. This panel does not behave like an Euler column, because the term $D_x L^2 / B^2$ is significant, but rather it is like an orthotropic plate with simple supports along each side.

5.4 FAILURE THEORIES OF STIFFENED PLATE PANELS WITH ELEMENTS OF HIGHER b/t RATIO

When a stiffened plate buckles as a wide column towards the plate (Fig. 5.2) the tip of the stiffener carries the highest compressive stress. Provided the stiffeners are sufficiently stocky so that they do not buckle prematurely the same criterion of failure as that used in Section 5.2 may again be employed here, i.e., failure is assumed to occur when the maximum compressive stress reaches the yield stress. When failure is in the opposite direction and the plate elements carry the greatest compressive stress and if their b/t ratio is sufficiently large they will buckle before the maximum axial load is attained. This section is concerned with the second of these two methods of failure.

When the b/t ratio of one or more of the plate elements which form the cross-section exceeds 45, local elastic buckling becomes an important initial part of the failure process. This is so because local elastic buckling leads to the development of a local plastic mechanism which results in a global failure mechanism. It is well known that a flat square plate $b \times b \times t$ with simply-supported sides and an axial compressive stress σ acting in one direction only buckles elastically when σ reaches

$$\sigma_{cr} = \frac{4\pi^2 E}{12(1 - v^2)} \left(\frac{t}{b}\right)^2 \tag{5.23}$$

Designers like to use elements with b/t values in the range 40–55 because they feel that they are the most efficient b/t ratios.

For mild steel whose yield stress is 250 MPa the b/t ratio at which σ_{cr} is equal to the yield stress is

$$\frac{b}{t} = \left[\frac{4\pi^2 \times 20\,600}{12(1 - 0 \cdot 3^2)} \times \frac{1}{250}\right]^{1/2} = 54$$

and for steel with a yield stress of 350 MPa the corresponding b/t ratio is 46. In practice it is found that local buckling must be considered if an element b/t ratio exceeds 35. The length of the panel is unimportant when considering local buckling because the elements try to buckle into square patterns. Thus a simply-supported plate $L \times b \times t$ with $L = \alpha b$ where α is an integer, will buckle into α squares and the buckling stress is given by eqn (5.23). When α is not an integer the buckling stress is approximately equal to the value given in eqn (5.23), provided $\alpha > 1$. Figure 5.5(a) shows such a plate with an aspect ratio (L/b) of 3·8 and the buckling pattern. The distribution of axial stress before and after buckling is also shown and it is seen that there is a concentration of stress towards the edges and a reduction of stress in a central strip of plate. Figure 5.5(b) shows the graphs of maximum deflection and average axial strain against the applied in-plane stress for elastic perfectly-flat plates and plates with initial imperfections. These and other results have been obtained (Coan, 1951; Stein, 1959; Yamaki, 1959; Walker, 1969) by solving Marguerre's equations (Section 5.3). It is seen that after buckling elastically these plates can carry further increases in applied stress but that at the point of buckling

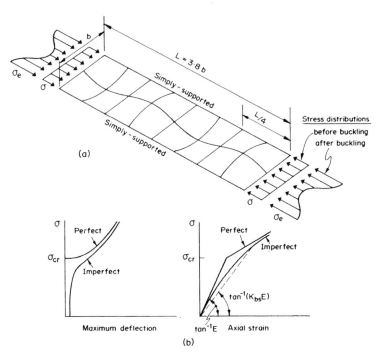

FIG. 5.5.   (a) A long simply-supported plate buckles into nearly square panels with a redistribution of stress after buckling. (b) Curves of maximum deflection and axial strain against applied stress for perfectly flat and imperfect plates.

the axial stiffness will decrease. Thus although post-buckling strength is available, it is at the cost of considerable loss of stiffness.

When a long strip of plate forms one element of the cross-section of a stiffened plate panel, its conditions along the lateral edges are usually not simply-supported. If the strip is more flexible than adjacent elements to which it is joined, its buckling stress will exceed the value given by eqn (5.23) because it is partially restrained against rotation at its edges by the other elements. If it is stiffer than the surrounding elements, its buckling stress will be less than the value given by eqn (5.23). In fact, because the elements must all buckle simultaneously, we should think only of the critical stress for the whole cross-section. The critical stress can be evaluated by a method developed by Williams and Wittrick (1969) or by the finite strip method of Cheung (1976). However, extensive tables of critical stresses for a wide range of cross-sections have been prepared by Klöppel and Scheer (1960), Klöppel and Möller (1968) and Murray and Thierauf (1981). The work of the last reference is explained in more detail in Thierauf et al. (1982) and Murray (1983).

The reader will find that in many textbooks on plate buckling the critical stress is expressed as a modified form of eqn (5.23), viz.,

$$\sigma_{cr} = \frac{k\pi^2 E}{12(1 - v^2)} \left(\frac{t}{b}\right)^2 \tag{5.24}$$

where k is a dimensionless factor which takes account of the boundary conditions along the edges of the plate.

The problem dealt with in this section is to estimate the failure load of a stiffened plate panel whose plate elements have buckled locally and elastically in the manner just described. The following is mainly a description of the phenomenon of collapse with an indication of the underlying theoretical concepts. A more comprehensive description is available from the source papers listed in the references and a detailed review of a number of failure theories with background theory is given by Murray (1983). This section simply deals with three quite different philosophical approaches to the problem of evaluating the collapse load of a stiffened plate panel.

The simplest theories are empirically based upon data collected from laboratory and other tests. Various formulae have been proposed, a typical example being that by Allen (1975):

$$\sigma_m^{-n} = \sigma_{cr}^{-n} + \sigma_E^{-n} + \sigma_y^{-n} \tag{5.25}$$

where

$n = $ constant (2 is suggested by Allen, 1975),

$\sigma_m = $ average stress acting on the cross-section at failure,

$\sigma_{cr} = $ average stress when local buckling occurs,

$\sigma_E = $ Euler stress of panel acting as a column $(= \pi^2 E(L/r^2)^{-2})$,

$\sigma_y = $ yield stress.

This equation is a modified Rankine formula. Rankine's equation for the failure stress of a pin-ended column is

$$\sigma_m^{-1} = \sigma_E^{-1} + \sigma_y^{-1} \qquad (5.26)$$

It is seen that eqn (5.25) is an extension of eqn (5.26) to take account of the interactive effects of local buckling, global buckling and yielding. Studies (Murray, 1983) show that over a wide range of results eqn (5.25) gives reasonably good results, but for some individual panels it gave remarkably poor results. One reason for this is that it gives the absurd result that σ_m cannot exceed σ_{cr} whereas for higher b/t ratios theoretical and experimental evidence is to the contrary. Other empirical formulae (Herzog, 1976; de George, 1979) try to correct this by applying a coefficient to the term which includes σ_{cr}.

The second approach to the evaluation of the failure load of a panel is to use an effective width of plate element instead of its actual width. As seen in Fig. 5.6, the strips of plate adjacent to the supported edges carry greater stresses than those at the centre of the plate after it has buckled. Thus, it is argued, the real plate element can be replaced by an equivalent plate of reduced, or so-called effective, width in which the stress is uniformly distributed.

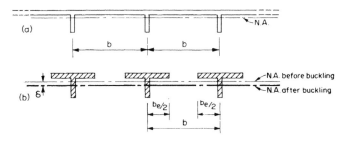

FIG. 5.6.   When a stiffened plate buckles locally in the plate its effective cross-section changes and the panel behaves like an eccentrically loaded column.

Let us consider the flat stiffened plate panel with pinned ends with the cross-section shown in Fig. 5.6(a). The axial stress is uniformly distributed over the cross-section. If the panel is sufficiently long it will buckle as a wide Euler column before it can buckle locally and then the method described in Section 5.2 can be applied. This will be possible if $\sigma_E \ll \sigma_{cr}$. An accurate value of σ_{cr} can be obtained from Murray and Thierauf (1981). Alternatively, if it can be guaranteed that the stiffeners are stocky and do not buckle, an approximate but conservative estimate is the critical stress of the plate element with the greatest b/t ratio in the cross-section. The test then reduces to seeing whether

$$L \gg \frac{br}{t} [3(1 - v^2)]^{1/2} \qquad (5.27)$$

where r is the radius of gyration of the cross-section of the panel.

If this test fails local buckling is important and at the point of buckling the effective cross-section suddenly changes from the original cross-section to that shown shaded in Fig. 5.6(b). The latter cross-section is referred to as the effective cross-section and it is seen that the shift δ in the neutral axis means that after buckling the panel behaves like an eccentrically loaded column because the axial load is still applied at the original neutral axis. Now let us suppose that the original panel had an initial bow of δ_0 over its span L in a direction which results in additional bending compressive stresses in the plate; then as a simple approximation the panel can be treated as a wide pin-ended column with initial bow $\delta + \delta_0$. The section properties of the effective column are derived from the effective cross-section. The effective column is then analysed by the method presented in Section 5.2 and the resulting equation is another modified version of the Perry–Robertson formula. This is a simplified description of how this type of panel has been treated by Horne and Narayanan (1975, 1976) and by Murray (1975) although these papers differ somewhat in their details.

If the lateral edges of the panel are simply supported, the method presented in Section 5.3 which treats the panel as an orthotropic plate can be modified in a similar way.

A description of this method would not be complete without a summary of how the effective width $b_e^{'}$ of a given plate element of width b is defined. The following describes but one of the many methods which are available for estimating b_e. Let the stress at the edges of the buckled element be σ_e and let σ_{av} be the average stress over the whole plate element. It is assumed that the real element can be replaced by an equivalent element which consists of

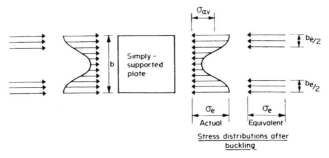

FIG. 5.7. The non-uniform stress distribution in a plate after buckling is replaced by an equivalent distribution.

two blocks of plate (Fig. 5.7) each of which carries a uniform stress σ_e. Equating forces in the real and equivalent plates

$$\sigma_{av} bt = \sigma_e b_e t \qquad (5.28)$$

i.e.

$$b_e/b = \sigma_{av}/\sigma_e = K_{bs} \qquad (5.29)$$

where K_{bs} is called the secant effective width factor. It is easily shown (Murray, 1983) that it is also the factor by which the axial stiffness of the original element is reduced as a result of buckling [Fig. 5.5(b)]. By solving Marguerre's equations (see Section 5.3) for a simply-supported plate with initial imperfections at the centre of y_0 it can be shown (Murray, 1983) that the value of K_{bs} is given by the equation:

$$K_{bs} = \frac{1 + 2m(m_1 + 1)C}{1 + 4m(m_1 + 1)C} \qquad (5.30)$$

where

$$m_1 = (w_0 + y_0)/y_0 \qquad (5.31)$$

$w_0 + y_0$ = total central deflection

$$C = \frac{3(1 - v^2)}{16} \left(\frac{y_0}{t}\right)^2 \qquad (5.32)$$

If it is assumed that failure occurs when σ_e reaches the yield stress σ_y the secant effective width factor is

$$K_{bs} = 1 - \frac{(m_1^2 - 1)\pi^2}{8} \left(\frac{E}{\sigma_y}\right) \left(\frac{y_0}{t}\right)^2 \left(\frac{t}{b}\right)^2 \qquad (5.33)$$

By eliminating m_1 between eqns (5.30) and (5.33), K_{bs} can be expressed as a function of y_0/t and $(b/t)(\sigma_y/E)^{1/2}$. These functions have been plotted as design curves by Horne and Narayanan (1975) and they are also available in Murray (1983).

The method just described appears to give reliable results when gauged against experimental data but it has been criticised by Dwight and Little (1976) and Little (1976). The basis of their criticism is that effective width methods are based upon the assumption that local buckling always occurs whereas it is well-known that for high L/r ratios local buckling may not occur until after the process of global collapse is well developed. They claim that an effective stress method is more rational and reliable because it uses the actual load-shortening curves of plate elements. The following is a brief description of their method.

The first stage of their investigations was carried out by Moxham (1971*a*, *b*) who studied the theoretical and experimental behaviour of an isolated plate with initial imperfection and uniaxial stress. The theoretical work involved the division of the plate into elements whose shape approximates that of a rectangular parallelepiped. There were five such elements through the thickness and 18 in both the longitudinal and transverse directions. In the analysis, which incidentally is not a finite element analysis, it was assumed that the shape of the deflected plate can be expressed as a Fourier series and after each load increment the coefficients of the series were calculated so as to minimise the strain energy. The plastic effects were catered for by using the Prandtl–Reuss flow rule and the von Mises criterion. Figure 5.8, which is taken from some later work of Little (1980), shows three typical graphs of axial stress σ_{av} against axial strain ε for

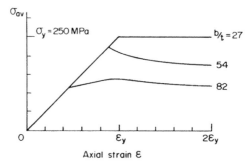

FIG. 5.8.   Stress–strain curves of three perfect plates showing the effect of b/t ratio (Little, 1980).

plates with no initial deflection and no residual stresses. It is seen that when $b/t = 27$ the plate behaves in a ductile manner with a well-defined yield plateau. When $b/t = 82$ the graph has two slopes before the buckling plateau is developed. In the first loading stage the plate is in its pre-buckled state and in the second, elastic buckling has reduced the axial stiffness of the plate because of the redistribution of stresses. In the third phase yielding is the dominant effect. When $b/t = 54$, failure occurs at a high stress but there is a rapid unloading of the plate after this point. It was shown in Section 5.4 that plates with b/t values in the range 45–55 are often used in thin-walled structures because their critical stress and yield stress are about equal and therefore at first sight they appear to be economic. Studies by Moxham (1971a, b) and others on plates with initial imperfections show that the effect of the initial imperfections is to round out the discontinuities in the graphs such as those shown in Fig. 5.9. Furthermore studies on the effects of residual stresses showed that they cause yielding to occur earlier and may result in considerable reduction in stiffness. Thus the high failure stress of perfect plates with b/t values in the intermediate range of 45–55 appears to be an unattainable or at least unreliable goal in practice.

Moxham's results, which have recently been extended by Little (1980) and by Harding et $al.$ (1977a, b) and Harding and Hobbs (1979), were applied by Dwight and Little (1976) and Little (1976) to the problem of finding the collapse load of a stiffened plate panel with simply-supported sides and uniaxial compression. They have produced the following simple method which is suitable for design purposes. They treated the panel as a wide column and followed the strain history of each element within it. The interactive effects of local and global buckling are considered by first using design curves (Fig. 5.9(a)) which determine for each plate element the

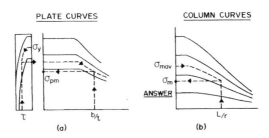

Fig. 5.9.   The average σ_m at failure of a stiffened panel is found: (a) by determining the failure stress σ_{pm} of each plate element; (b) by averaging to find σ_{mav} and using the column curves as shown (Dwight and Little, 1976).

average stress at failure σ_{pm} as a function of b/t. The nomogram on the left enables the effects of in-plane shear stress to be included. These stresses are then averaged over the whole cross-section to obtain $\sigma_{m\,av}$. The value of $\sigma_{m\,av}$ is then used in a column curve of average stress at failure as a function of L/r, i.e. a Perry–Robertson type of curve (Fig. 5.9(b)). Allowances can be made for different classes of weld and for the magnitude of the initial imperfections of the column. In this method buckling of the stiffeners is avoided by requiring the designer to comply with certain rules about their depth-to-thickness ratio. Their work has been extended to stiffened panels in box-columns (Little, 1979).

The methods described here and several others have been tested on some experimental results obtained on six very large box girders by Mikami *et al.* (1980) and on a smaller box girder by Roderick and Ings (1977). However, there have been few other attempts to analyse a stiffened plate panel when it is a part of a structure. Maquoi and Massonnet (1971) developed a method which assumes that the compression flange of a box girder behaves like an orthotropic plate. Failure was assumed to occur when the membrane stress at each of the lateral edges of the flange reached the yield stress. At present there is a need to test more stiffened plates of different geometries to confirm the accuracy and reliability of the available theories. Also the behaviour of plates with other than rectangular stiffeners has not been thoroughly investigated.

5.5 BEHAVIOUR OF STIFFENED PLATES UNDER COMBINED AXIAL AND NORMAL LOADING

Interesting investigations were carried out by Michelutti (1976) and de George *et al.* (1979) and also by de George (1979) on the behaviour of stiffened plates under combined axial and normal loading. Michelutti tested 14 full-scale panels each being nominally identical with the others. The stiffeners were bulb-flats. An interaction curve at failure of axial load against transverse bending moment was obtained [Fig. 5.10(a)]. It was shown that when the transverse loading was arranged so that the tip of the stiffener was in compression the criterion of failure which gave reliable results was that of first yield at that point. When the loading was in the opposite direction, two principal failure mechanisms occurred. Between these two mechanisms there was a sudden reduction in strength as is seen in the upper part of the diagram shown in Fig. 5.10(a).

De George carried out combined loading tests on 11 identical models of

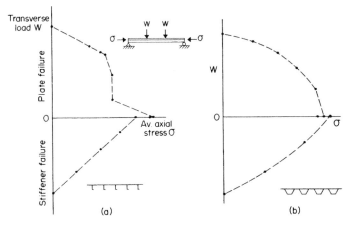

FIG. 5.10.   Interaction diagrams for failure of stiffened panels with combined axial and transverse loading (Michelutti, 1976; de George, 1979; de George *et al.*, 1979).

plates with trough stiffeners and showed that this form of panel does not have the same problems. The panels generally behaved like stocky beam-columns [Fig. 5.10(b)].

In both series of tests adequate rules for the design of these types of panel under combined loadings were put forward.

5.6   STUDIES OF BEHAVIOUR OF STIFFENED PANELS AFTER FAILURE

When a stiffened plate carries an increasing axial stress σ its behaviour is at first elastic, i.e. if σ is removed the plate will return to its unloaded state. Eventually σ is large enough to cause some yielding of the panel and it is then said to be in an elasto-plastic state. Finally, after the maximum value of σ has been attained a plastic collapse mechanism is formed and the load-carrying capacity of the panel diminishes. It was pointed out in Section 5.1 that the rate at which a given panel unloads after collapse is an important consideration for a designer. An approximate unloading curve can be developed by applying rigid-plastic theory (Murray, 1973*a*, *b*) to the localised spatial mechanisms which form usually near the centre of the panel. Figure 5.11 shows two alternative local mechanisms associated with a given panel with rectangular stiffeners.

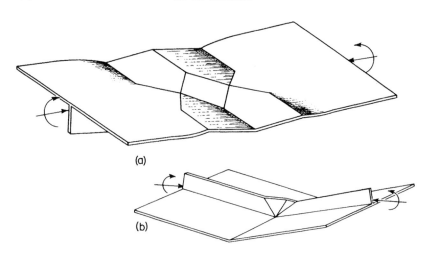

FIG. 5.11.   A simple stiffened plate panel can fail with (a) a plate mechanism or (b) a stiffener mechanism.

In order to study these mechanisms in a quantitative manner, it is necessary to appreciate the following points.

(a)   Some local mechanisms consist of plastic hinges only, while others require areas of the plate to yield throughout their thickness so that the mechanism can deform. The former are called true plastic mechanisms while the latter are called quasi-plastic mechanisms (Murray and Khoo, 1981).

(b)   The effect of an axial stress σ upon the plastic moment capacity per unit width of plate is

$$M'_p = \frac{\sigma_y t^2}{4} \, [1 - (\sigma/\sigma_y)^2] \qquad (5.34)$$

(c)   The moment capacity of a plastic hinge which is inclined at an angle β to the cross-section of the panel is proportional to $\sec^2 \beta$. Thus, the plastic moment capacity per unit width of plate when the hinge is so inclined is (Murray, 1973c)

$$M''_p = M'_p \sec^2 \beta \qquad (5.35)$$

Although local plastic mechanisms appear at first sight to be somewhat complicated they can usually be considered as an assembly of so-called basic mechanisms which work together in a compatible fashion.

A table of the governing equations of a number of basic mechanisms is now available (Murray and Khoo, 1981; Murray, 1983) and it enables most of the more complicated mechanisms to be analysed and their collapse curves to be described.

From these analyses it has been possible to confirm certain laboratory observations. For example, the plate failure mechanism shown in Fig. 5.11(a) is normally much more ductile than the stiffener mechanism shown in Fig. 5.11(b). Murray and Khoo (1981) have also analysed the local mechanisms of thin-walled I-beams, channels and box columns and it was shown that in some circumstances collapse can be very sudden. Furthermore, they show that simple plastic theory which ignores the possibility of local plastic mechanisms can greatly overestimate the load-carrying capacity in the post-buckling region.

5.7 TREATMENT OF THE PROBLEMS OF THIN-WALLED STRUCTURES BY SOME CODES OF PRACTICE

A review of some clauses of a few codes of practice has recently been published (Murray, 1983). In this last section of this chapter a very brief summary of a few examples only can be treated.

The problem facing the designer, viz., that local plate buckling and global column buckling can interact, is probably best illustrated by the German Code (DAst. Richtlinie 012, 1978). An interaction chart is shown diagrammatically in Fig. 5.12. The coordinates $(L/r, b/t)$ of a point in it define the average stress at failure of a panel. The abscissa is related to the buckling of the panel as a wide column with the L/r value being the predominant variable. The ordinate of the point is defined by the stress at which local buckling occurs. For the latter figure the designer is referred to published data on the buckling stresses of stiffened plates (Klöppel and Scheer, 1960; Klöppel and Möller, 1968; Murray and Thierauf, 1981). Having established the coordinates of the point the failure stress is read off the contour lines which run around the diagram. The dotted lines indicate how the $0 \cdot 6\sigma_y$ contour is established from the plate and column-buckling curves. The diagram has the advantage that it is easy to use and that it brings out the importance of the local and global buckling which can occur interactively in Region 3 of the diagram. One feature of this diagram is that for high b/t ratios failure is said to occur when the plate first buckles elastically. This is a conservative approach because it ignores the post-buckling strength of such plates (see Fig. 5.5(b)).

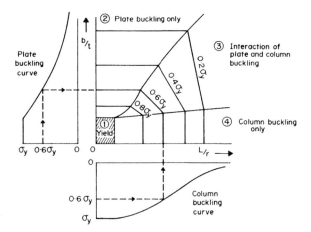

FIG. 5.12.   German design chart for failure stress of a stiffened plate panel with
axial stress (DAst Richtlinie 012, 1978).

The method due to Dwight and Little (1976) and Little (1979) for the
design of stiffened plate panels (Section 5.4) is another method which
clearly illustrates how the interactive effects of local and global buckling
may be treated. It has been incorporated into the British Code BS 2573
(1977) for the design of the box columns in cranes.

In the British bridge code BS 5400 (1978–81) it is recognised that the
failure of a single thin rectangular stiffener can cause sudden collapse of the
whole panel. In order to avoid this dangerous situation the dimensions of
the stiffeners are constrained. The average stress at failure σ_m is determined
by using an effective width method—appropriate graphs for the
determination of b_e and of σ_m as a function of L/r being provided for this
purpose.

When the in-plane stress acting on a plate element (e.g. in the web of a
box girder) is a combination of axial, lateral, bending and shear stresses,
some codes (e.g. Merrison Rules, 1973) suggest a method of determining
the equivalent elastic buckling stress. The concept employed is that for
design purposes a complex combination of stresses can be replaced by an
equivalent single axial stress which will eventually reach an elastic critical
value $\sigma_{e\,cr}$ as the loads are increased. Methods available for the
determination of $\sigma_{e\,cr}$ are described in Murray (1983). After this stress has
been evaluated the design procedure is the same as for a plate element with
an axial stress acting alone.

REFERENCES

AALAMI, B. and CHAPMAN, J. C. (1969) Large deflection behaviour of rectangular orthotropic plates, under transverse and in-plane loads. *Proc. ICE*, **42**, 347–82.

ALLEN, D. (1975) Discussion on Murray (1975). *Structural Engineer*, **53**, 381–2.

BASU, A. K. and CHAPMAN, J. C. (1966) Large deflection behaviour of transversely loaded rectangular orthotropic plates. *Proc. ICE*, **35**, 79–110.

BRITISH STANDARDS INSTITUTION (1977) *Specification for Permissible Stresses in Cranes and Design Rules*, BS 2573, Part 1.

BRITISH STANDARDS INSTITUTION (1978–81). *Steel, Concrete and Composite Bridges*, BS 5400.

CHEUNG, Y. K. (1976) *Finite Strip Method in Structural Analysis*, Pergamon Press, Oxford.

COAN, J. M. (1951) Large deflection theory for plates with small initial curvatures loaded in edge compression. *J. Appl. Mech.*, **18**; *Trans ASME*, **73**, 143–51.

DAST. RICHTLINIE 012 (1978) Beulsicherheitsnachweise für Platten. Publ. of Working Party PLATTEN of Deutsche Ausschuss für Stahlbau.

DE GEORGE, D. (1979) Collapse behaviour of trough stiffened steel plates. PhD Thesis, Monash University.

DE GEORGE, D., MICHELUTTI, W. M. and MURRAY, N. W. (1979) Studies of some steel plates stiffened with bulb-flat or with troughs. *Thin-walled Structures Conference*, University of Strathclyde, Granada Publications, pp. 86–9.

DWIGHT, J. B. and LITTLE, G. H. (1976) Stiffened steel compression flanges—a simpler approach. *Structural Engineer*, **54**(12), 501–9.

HARDING, J. E. and HOBBS, R. E. (1979) The ultimate load behaviour of box girder web panels. *Structural Engineer*, **57B**(3), 49–54.

HARDING, J. E., HOBBS, R. E. and NEAL, B. G. (1977a) Ultimate load behaviour of plates under combined direct and shear in-plane loading. *Steel Plated Structures Symposium* (Ed. Dowling, Harding and Frieze), Crosby and Lockwood, London, pp. 369–403.

HARDING, J. E., HOBBS, R. E. and NEAL, B. G. (1977b). The elasto-plastic analysis of imperfect square plates under in-plane loading. *Proc. ICE*, **63**(2), Paper No. 7981, 137–58.

HERZOG, M. (1976). Die Traglast einseitig längsversteifter Bleche mit Imperfektionen und Eigenspannungen unter Axialdruck nach Versuchen. *VDI-Z*, **118**(7), 321–6.

HORNE, M. R. and NARAYANAN, R. (1975) An approximate method for the design of stiffened steel compression panels. Manchester University Simon Eng. Lab. Report.

HORNE, M. R. and NARAYANAN, R. (1976) Ultimate capacity of longitudinally stiffened plates used in box girders. *Proc. ICE*, **61**(2), 253–80.

KLÖPPEL, E. K. and MÖLLER, K. H. (1968) *Beulwerte ausgesteifter Rechteckplatten*, Vol. II, W. Ernst, Berlin.

KLÖPPEL, E. K. and SCHEER, J. (1960) *Beulwerte ausgesteifter Rechteckplatten*, Vol. I, W. Ernst, Berlin.

LITTLE, G. H. (1976) Stiffened steel compression panels—theoretical failure analysis. *Structural Engineer*, **54**(12), 489–500.

LITTLE, G. H. (1979) The strength of square steel box columns—design curves and their theoretical basis. *Structural Engineer*, **57A**(2), 49–61.

LITTLE, G. H. (1980) The collapse of rectangular steel plates under uniaxial compression. *Structural Engineer*, **58B**(3), 45–61.

MAQUOI, R. and MASSONNET, C. (1971) Theorie non-lineaire de la resistance postcritique des grandes poutres en caisson raidies. *Int. Ass. of Bridge and Struct. Eng. Pub.*, **31**(2), 91–140.

MERRISON RULES (1973) *Inquiry into the Basis of Design and Method of Erection of Steel Box Girder Bridges*, HMSO, London.

MICHELUTTI, W. M. (1976) Stiffened plates in combined loading. PhD Thesis, Monash University.

MICHELUTTI, W. M. and MURRAY, N. W. (1977) The collapse behaviour of stiffened plates under combined axial and bending loads. *Proc. 6th Australasian Conf. on the Mech. of Struct. and Materials*, Christchurch, New Zealand.

MIKAMI, I., DOGAKI, M. and YONEZAWA, H. (1980) Ultimate load tests on multi-stiffened steel box girders. Technology Reports of Kansai University, Osaka, Japan, No. 2, 157–69.

MOXHAM, K. E. (1971a) Theoretical prediction of the strength of welded steel plates in compression. Cambridge University Dept. of Engineering Report, CUED/C-Struct/TR2.

MOXHAM, K. E. (1971b) Buckling tests on individual welded steel plates in compression. Cambridge University Dept. of Engineering Report, CUED/C-Struct/TR3.

MURRAY, N. W. (1973a) Buckling of stiffened panels loaded axially and in bending, *Structural Engineer*, **51**, 285–301.

MURRAY, N. W. (1973b) Das Stabiltätsverhalten von axial belasteten, in der Längsrichtung ausgesteiften Platten im plastischen Bereich. *Der Stahlbau*, **12**, 372–9.

MURRAY, N. W. (1973c) Das aufnehmbare Moment in einem zu Richtung der Normalkraft schräg liegenden plastichen Gelenk. *Die Bautechnik*, **2**, 57–8.

MURRAY, N. W. (1975) Analysis and design of stiffened plates for collapse load. *Structural Engineer*, **53**, 153–8.

MURRAY, N. W. (1983) *Introduction to the Theory of Thin-walled Structures*, Oxford University Press, Oxford.

MURRAY, N. W. and KHOO, P. S. (1981) Some basic plastic mechanisms in thin-walled steel structures. *Int. J. Mech. Sci.*, **23**(12), 703–13.

MURRAY, N. W. and THIERAUF, G. (1981) *Tables for the Design and Analysis of Stiffened Steel Plates*, Vieweg, Braunschweig.

PARKES, E. W. (1965) *Braced Frameworks*, Pergamon Press, Oxford.

RODERICK, J. W. and INGS, N. L. (1977) The behaviour of small scale box girders of stiffened plate construction. *Australian Welding Research*, pp. 16–29.

ROSTOVTSEV, G. G. (1940) Calculation of a thin-plate sheeting supported by rods. Londy, Leningrad Inst., Inzhenerov, Grazhdanskogo Vasdushnogo Flota, No. 20.

STEIN, M. (1959) Load and deformations of buckled rectangular plates. NASA Tech. Report R-40.

THIERAUF, G., KATZER, W. and MURRAY, N. W. (1982) Application of the finite strip method to the design of stiffened plates for buckling. *International Conf. on Finite Element Methods*, Shanghai, China.

WALKER, A. C. (1969) The post-buckling behaviour of simply-supported square plates. *Aeronautical Quart.*, **XX**, 203–22.

WILLIAMS, F. W. and WITTRICK, W. H. (1969) Computational procedures for a matrix analysis of the stability and vibration of thin flat-walled structures in compression. *Int. J. Mech. Sci.*, **11**, 979–88.

YAMAKI, N. (1959) Post-buckling behaviour of rectangular plates with small initial curvatures loaded in edge compression. *J. Appl. Mech.*, **26**, 407–14.

Chapter 6

SHEAR LAG IN BOX GIRDERS

V. Křístek

Faculty of Civil Engineering, Czech Technical University,
Prague, Czechoslovakia

SUMMARY

The shear lag phenomenon in steel box girder bridges is studied. Various methods of the shear lag analysis are presented and compared. The influence of various parameters such as flange geometry and orthotropy, initial imperfections, support and loading conditions and effects of longitudinal and transverse stiffeners are examined; shear lag in curved bridges and its effect on the distribution of the longitudinal stresses across the flange width are studied. Suggestions to reduce adverse effects of shear lag are made, and recommendations for the practical design are presented.

6.1 INTRODUCTION

In a box girder (Fig. 6.1(a)) the web and flange plates are interconnected so that relative displacements cannot occur between them. Therefore at the junction of the web with the flange, the longitudinal strain in the web ($\varepsilon_{x,w}$) must be equal to that in the flange ($\varepsilon_{x,f}$). A shear flow is developed between the web and the flange which causes shear deformation of the flange plate. The longitudinal displacements in the parts of the flange remote from the webs lag behind those nearer the webs. This effect leads to a non-uniform distribution of the longitudinal normal stresses across the flange width. The effect is particularly pronounced in wide flanges and in flanges with longitudinal stiffeners.

V. KŘÍSTEK

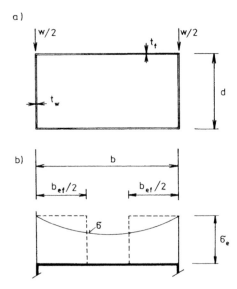

FIG. 6.1.   (a) Shear lag in a box girder. (b) The effective breadth concept.

This phenomenon termed shear lag results in a considerable increase of the longitudinal stresses σ in the regions of the flange close to the webs in comparison with those given by the elementary theory of bending (Fig. 6.1(b)). Thus, the neglect of shear lag would lead to an underestimation of the stresses developed in the flange plates at positions adjacent to the webs, and hence, to an unsafe design. The shear lag may also significantly influence the girder deflections.

It is an accepted practice in structural engineering to represent the effect of shear lag by adopting an effective breadth concept. The actual width of the flange plate b is replaced by a reduced width b_{ef} over which the longitudinal stresses may be considered uniformly distributed and the application of the elementary theory of bending to the transformed girder cross-section gives the correct value of a maximum longitudinal stress σ_e (Fig. 6.1(b)). A similar procedure may also be carried out for deflections. However, when the structure is subjected to large concentrated loads, the concept of effective breadth gives reliable information only in those parts of the structure that are not very close to the point of application of the load or the support reaction. In the immediate neighbourhood of the point load the actual stress state can differ rather substantially from that resulting from any simplified analysis, including the effective breadth concept.

6.2   METHODS OF ANALYSIS

Extensive work has been carried out to analyse the shear lag effect during the last few decades. An analytical method has been presented by Girkmann (1954). More recently many analytical models and methods have been developed. Among these are numerical solutions based on Finite element or Finite difference methods, exact and approximate methods based on Folded plate theory, and approximate methods based on simplified structural behaviour. As these methods are well-documented in the literature no attempt will be made to review all of them in detail here.

6.2.1   The Finite Element Method
The Finite element method has become a practically universal method for the solution of mechanics problems in recent years. The continuum is replaced by an assembly of finite elements interconnected at nodal points. Stiffness matrices are developed for the finite elements based on assumed displacement patterns and subsequently an analysis based on the direct stiffness method may be performed to determine nodal point displacements and subsequently the internal stresses in the finite elements.

Moffat and Dowling (1975) produced a comprehensive parametric study of the shear lag effect in box girders; this study was based on the use of the finite element method. They found that while only one mesh division over the depth of a girder is sufficient the fine mesh divisions must be used over the girder width and length, particularly in the region of a point load or a support. The finite element results obtained by Moffat and Dowling are generally accepted as a basis for the design rules (Dept of the Environment, 1973). These rules provide values which enable the effective flange width to be determined at all the positions along the span of a box girder of any plan dimensions and cross-sectional proportions and subjected to distributed or point loading.

Although a finite element solution is capable of giving a comprehensive and adequate picture of the stress distribution this solution requires the use of large computers and is too costly particularly if repeated analyses are required at the preliminary design stage.

6.2.2   The Folded Plate Theory
Steel box girders are usually of constant cross-section and hence the folded plate theory is an ideally suited technique (DeFries-Skene and Scordelis, 1964; Lin and Scordelis, 1971; Křístek, 1979a) for predicting shear lag effects. A prismatic folded plate structure (Fig. 6.2(a)) is a shell consisting

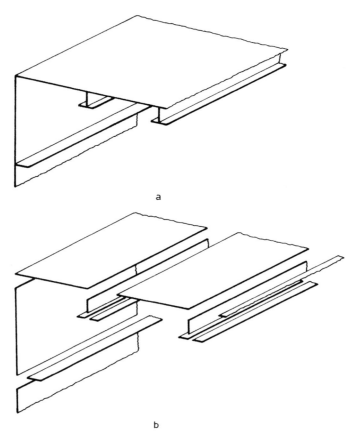

FIG. 6.2.   Folded plate structural idealisation.

of rectangular plates, mutually supporting along their longitudinal edges and simply-supported at the two ends by transverse diaphragms.

The folded plate analysis takes full advantage of the capabilities of digital computers. The complexity of the cross-section as well as the ratio of the cross-sectional dimensions and the span are irrelevant because the solution is not based on the elementary theory of bending but on the elastic plane stress theory and the plate bending theory. Bearing in mind the assumptions of the theory of elasticity the Folded plate theory is in fact an exact method because it considers the structure in its actual form as an assembly of plate elements forming together a real spatial system (Fig. 6.2(b)). The method has the advantage that it can treat any general set of joint loads or displacements.

Each folded plate element is subjected to element forces and displacements at its longitudinal edges. There are two edges; hence the plate has eight degrees of freedom, eight element forces and eight element displacements.

Due to the diaphragms (rigid in their own plane and perfectly flexible in the direction normal to it) at both ends of the folded plate structure, the distribution of all applied forces and displacements along a ridge or joint is taken as a harmonic of order n. This makes it possible to treat the ridge as a nodal point and to operate with single forces and displacements instead of functions. If the conditions of static equilibrium and geometric compatibility are satisfied at this nodal point they will be automatically maintained along the entire ridge; this results in a considerable saving in computer time.

As a first step of a solution, an element stiffness matrix (8×8) is written, relating the element forces in the relative system and the corresponding displacements. Then a displacement transformation matrix (8×8) is defined, relating the element displacements in the relative coordinate system and in the fixed system. Applying this transformation matrix the element stiffness matrix in the fixed coordinate system may be found. The stiffness matrix for the whole structure can now be assembled from the element stiffness matrices.

The objective of the general stiffness method is to find all unknown joint displacements. Once these are known it is possible to determine all internal forces and stresses acting in the structure.

The results of a typical numerical calculation are presented in Fig. 6.3 for a steel box girder previously analysed by Moffat and Dowling (1975). Figure 6.3 shows a comparison of results using both the Folded plate theory and the Finite element method. The study was conducted on a box girder having a width of 3658 mm, depth of 1829 mm, flange thickness of 25·4 mm and web thickness of 12·7 mm. The span of the girder was 9144 mm and subjected to uniformly distributed loading above the webs.

The agreement between the values obtained from the Finite element method (Moffat and Dowling, 1975) and those given by Křístek et al. (1981) using the Folded plate theory (in terms of the effective breadth ratios) is extremely good.

6.2.3 The Finite Strip Method

The direct application of the theory of elasticity to determine the stiffness matrix of folded plate elements which are curved or which are not isotropic becomes exceedingly complex. A theory known as the Finite strip method

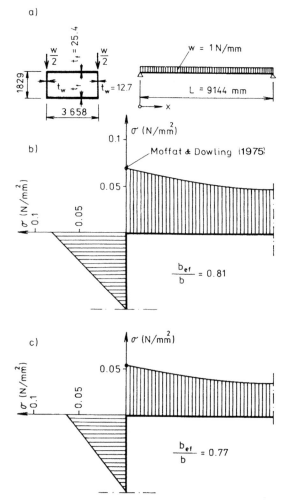

FIG. 6.3. Comparison of results of Folded plate and Finite element (Moffat and Dowling, 1975) methods.

(Meyer and Scordelis, 1970) may be used in these cases. This method may be considered as a special form of the Finite element method. It approximates the behaviour of each plate by an assembly of longitudinal finite strips for which selected displacement patterns varying as harmonics longitudinally and as polynomials in the transverse direction are assumed to represent the behaviour of the strip in the total structure. With this assumption the displacement at any point in the strip can be expressed in terms of eight nodal point displacements and hence the element stiffness matrix determined. The remaining procedure is similar to that used in the Folded plate method.

6.2.4 The Bar Simulation Method

Evans and Taherian (1977) proposed two simplified methods which enable the designer to calculate shear lag effects without having to refer to tabulated empirical values. The first procedure requires the use of a small computer program and the second method gives a slightly less accurate prediction from simple hand calculations. The methods are based on early work carried out in the aircraft engineering field (Kuhn, 1956).

It is proposed that in the Bar simulation method (Evans and Taherian, 1977) an unstiffened plate, such as that shown in Fig. 6.4(b), may be artificially divided into axial load-carrying and shear-carrying components, as shown in Fig. 6.4(c). The axial load-carrying capacity of the plate is assumed to be concentrated at a number of equally-spaced bar members of equal area. The area of each bar A is equal to the cross-sectional area of the flange plate (i.e. bt_f) divided by the total number of bars n (Fig. 6.4(c)). The sheet itself is then assumed to be capable of

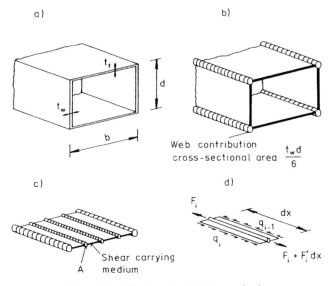

FIG. 6.4.   The bar simulation method.

carrying shear stresses only. The forces F applied to the bars must be in equilibrium with those applied to the sheet and the deformations of the bars and the sheet must be compatible.

Considering the equilibrium of the forces applied to a differential length of each bar (Fig. 6.4(d))

$$\frac{\mathrm{d}F_i(x)}{\mathrm{d}x} = q_{i-1}(x) - q_i(x) \tag{6.1}$$

in which q is a shear flow in the sheet.

By considering the shear deformations of the typical sheet segment i bounded by bars i and $i+1$, the shear strains γ_i in the sheet may be expressed in terms of the direct strains ε in the bars

$$\frac{\mathrm{d}\gamma_i}{\mathrm{d}x} = \frac{\varepsilon_{i+1} - \varepsilon_i}{b_f(n-1)} \tag{6.2}$$

Replacing the strains by stresses according to the relationships

$$\gamma_i = q_i/t_f G \qquad \varepsilon_i = \sigma_i/E \qquad \varepsilon_{i+1} = \sigma_{i+1}/E \tag{6.3}$$

where E is the modulus of elasticity of the bars and G is the shear modulus of the sheet, the shear flows in the sheet segments may be expressed in terms of the axial forces in the bars.

$$\frac{\mathrm{d}q_i}{\mathrm{d}x} = \frac{t_f G}{Eb_f/(n-1)} \left(\frac{F_{i+1}}{A_{i+1}} - \frac{F_i}{A_i} \right) \tag{6.4}$$

An equation such as eqn (6.4) may be written for each sheet segment. Thus a number of simultaneous differential equations can be established. The boundary conditions relating to F and q must be specified before these equations can be solved for the values of the axial forces F and the shear flows q.

The method is approximate because it involves the idealisation of a continuous plate by a series of discrete bar members. The accuracy depends on the number of bars taken but the computation effort and the solution time increases with the number of bars.

The bar simulation method (Evans and Taherian, 1977) requires the solution of a set of simultaneous differential equations, the number of equations being equal to the number of bars assumed in the idealisation. Such a solution can be most conveniently carried out on a digital computer. However, there are many occasions such as the initial estimation of girder proportions when the use of a computer is not justified. Therefore the simpler three-bar method (Evans and Taherian, 1977), which uses an empirical factor and enables a solution to be obtained by hand calculation, may be used instead of the bar simulation method with little loss of accuracy.

The three-bar method uses an approximate approach in which the system of seven discrete bars established in the bar simulation method is replaced by a three-bar system. No modifications of the properties of the edge bars are required. However, the other bars on the half-width of the plate are replaced by a single bar of equivalent area positioned at their centroid as defined by b_c. The half-width b_s of the substitute panel is then determined from the empirical relationship established by Kuhn (1956):

$$b_s = \left(0{\cdot}55 + \frac{0{\cdot}45}{m^2} \right) b_c \qquad (6.5)$$

where m is the number of bars between the edge and centre bars in the original system.

6.2.5 Harmonic Analysis of Shear Lag in Flanges with Closely-Spaced Stiffeners

A hand calculation of the shear lag effect in unstiffened flanges and in flanges with closely-spaced stiffeners was proposed by Křístek and Evans (1983). A simple method employing harmonic analysis is intended for use as a design tool.

A continuous girder with various support conditions can be approximated as an assembly of simply-supported beams and cantilevers, as shown in Fig. 6.5. The bending moment and shear force diagrams are established from continuous beam analysis in the first instance, as in Fig. 6.5(b), thereby defining the points at which the bending moment becomes zero and the shear force is known. The shear lag analysis is then carried out for each individual portion of the beam (see Fig. 6.5(c)). Harmonic analysis can now be applied directly to the case of simply-supported girders and cantilevers may be analysed by first establishing a substitute simply-supported beam (Křístek, 1979b; Křístek and Evans, 1983) as shown in Fig. 6.6.

It is assumed that the shear effects are transmitted by the flange plate (thickness t_f) itself but that the longitudinal axial forces are carried by both the stiffeners and the flange. Therefore it is necessary to introduce a modified flange thickness $\bar{t} = t_f + A_s/a$, as shown in Fig. 6.7(c) (A_s is the cross-sectional area of each stiffener and a is the stiffener spacing, as shown in Fig. 6.7(b)) for the axial force-carrying action.

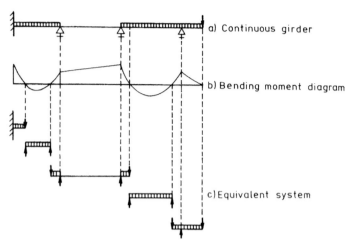

FIG. 6.5. Representation of a continuous girder by an assembly of simply-supported beams and cantilevers.

The method yields the following differential equation for the normal force n_x per unit width (Fig. 6.7(d))

$$\frac{\partial^2 n_x}{\partial y^2} + \left(\frac{\bar{t}E}{t_f G} - v\right)\frac{\partial^2 n_x}{\partial x^2} = 0 \tag{6.6}$$

where v is Poisson's ratio.

The solution may be expressed by the Fourier series

$$n_x(x, y) = \sum_{j=1}^{\infty} N_j(y)\sin\frac{j\pi x}{L} \tag{6.7}$$

where L is the length of the simply-supported span (Fig. 6.5(c)).

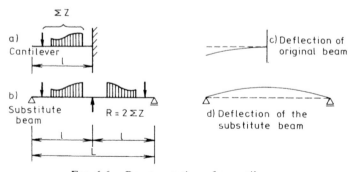

FIG. 6.6. Representation of a cantilever.

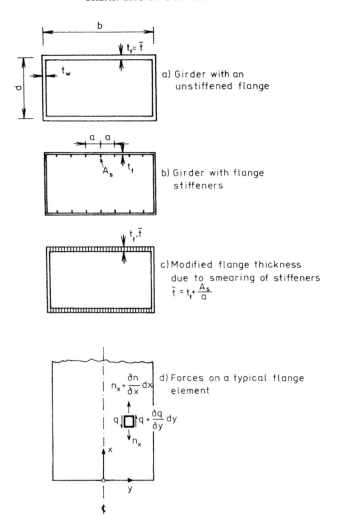

a) Girder with an unstiffened flange

b) Girder with flange stiffeners

c) Modified flange thickness due to smearing of stiffeners
$\bar{t} = t_f + \dfrac{A_s}{a}$

d) Forces on a typical flange element

FIG. 6.7.   Idealisation of flange plate.

Considering the j-th term of the series only, and substituting into eqn (6.6) the following ordinary differential equation is obtained

$$\frac{\mathrm{d}^2 N_j(y)}{\mathrm{d}y^2} - \frac{j^2 \pi^2}{L^2} \left(\frac{\bar{t}E}{t_f G} - v \right) N_j(y) = 0 \qquad (6.8)$$

Shear lag analysis is carried out for the loads that are placed symmetrically on the cross-section of a girder (Fig. 6.1(a)). Thus, assuming the origin of the transverse coordinate y to be taken at the mid-point of the flange (Fig. 6.7(d)), then due to symmetry, the distribution across the flange width of the normal force is governed by

$$N_j(y) = C_j \cosh \frac{j\pi y}{L} \sqrt{\left(\frac{\bar{\imath}E}{t_f G} - v\right)} \qquad (6.9)$$

The value of the constant C_j can be determined from the shear-loading condition at the edge of the plate.

From simple beam theory, the shear flow $q_e(x)$ transmitted from the web to the edge of the flange can be approximated as

$$q_e(x) = V(x) \frac{\bar{\imath} b d}{4I} \qquad (6.10)$$

where $V(x)$ is the total shear force acting on the beam cross-section at position x and I is the second moment of area of the complete cross-section, including stiffeners.

The shear flow transmitted at the edge can also be expressed in the form of a Fourier series. For the case of simply-supported ends, the series takes the form

$$q_e(x) = \sum_{j=1}^{\infty} Q_{e,j} \cos \frac{j\pi x}{L} \qquad (6.11)$$

where

$$Q_{e,j} = \frac{2}{L} \int_0^L q_e(x) \cos \frac{j\pi x}{L} \, dx = \frac{\bar{\imath} b d}{2IL} \int_0^L V(x) \cos \frac{j\pi x}{L} \, dx \qquad (6.12)$$

Combining the equations of equilibrium, eqns (6.7) and (6.9), the shear flow at the edge of the flange (where $y = b/2$) can also be expressed as

$$q_e(x) = -\left(\frac{\bar{\imath}E}{t_f G} - v\right)^{-1/2} \sum_{j=1}^{\infty} C_j \sinh \frac{j\pi b}{2L} \sqrt{\left(\frac{\bar{\imath}E}{t_f G} - v\right)} \cos \frac{j\pi x}{L} \qquad (6.13)$$

By equating the two expressions for $q_e(x)$ from eqns (6.11) and (6.13), the constant C_j is obtained as

$$C_j = -\left(\frac{\bar{\imath}E}{t_f G} - v\right)^{1/2} \frac{Q_{e,j}}{\sinh \dfrac{j\pi b}{2L} \sqrt{\left(\dfrac{\bar{\imath}E}{t_f G} - v\right)}} \qquad (6.14)$$

Having thus determined the constant, the amplitude of the normal longitudinal force for any particular harmonic can be obtained from eqn (6.9) as

$$
N_j(y) = -\left(\frac{\bar{\imath}E}{t_f G} - v\right)^{1/2} Q_{e,j} \frac{\cosh\dfrac{j\pi y}{L}\sqrt{\left(\dfrac{\bar{\imath}E}{t_f G} - v\right)}}{\sinh\dfrac{j\pi b}{2L}\sqrt{\left(\dfrac{\bar{\imath}E}{t_f G} - v\right)}} \tag{6.15}
$$

The magnitude of this force varies across the width of the flange; its peak value occurs at the edge, i.e. where $y = b/2$

$$
N_j(b/2) = -\left(\frac{\bar{\imath}E}{t_f G} - v\right)^{1/2} Q_{e,j} \operatorname{cotanh}\frac{j\pi b}{2L}\sqrt{\left(\frac{\bar{\imath}E}{t_f G} - v\right)} \tag{6.16}
$$

Knowing the amplitude N_j, the value of the longitudinal normal force per unit width $n_x(x, y)$ may be determined from eqn (6.7) for any position of the flange. Also the shear flow at any point $q(x, y)$ may be determined from eqn (6.13) to complete the solution.

Although the method is suitable for hand calculation it has been programmed for a pocket calculator for added convenience (Křístek and Evans, 1983).

6.2.6 Harmonic Analysis of Shear Lag in Flanges with Widely-Spaced Large Stiffeners

A method developed by Evans and Křístek (1983) deals with the case of a flange plate stiffened by a few irregularly or widely-spaced large stiffeners. The method can take an actual arrangement of stiffeners into account. The structural idealisation employed is similar to that of the bar simulation method proposed by Evans and Taherian (1977, 1980) and Taherian and Evans (1977). The greatest advantage of this approach is that by employing harmonic analysis the shear lag effects can be predicted directly from hand calculations. In the previous method (Evans and Taherian, 1977; Taherian and Evans, 1977), the solution of simultaneous differential equations (eqns (6.4)) by computer was required unless the use of certain empirical relationships was made.

As in the bar simulation method, it is assumed that the axial load-carrying capacity of the stiffened flange is concentrated at a number of longitudinal bar elements situated at the stiffener positions. The flange sheet itself is assumed to be capable of carrying shear stresses only. The equivalent area $\bar{A}$ of each bar is taken as the cross-sectional area of the stiffener A_s, together with the area of the adjacent flange sheet of the thickness t_f, i.e.

$$\bar{A} = A_s + t_f a \qquad (6.17)$$

where a is the stiffener spacing.

From the compatibility of longitudinal displacements the difference in the longitudinal strains in two adjacent bars must be equal to the total change in strain of sheet connecting both bars. Thus, for bars i and $i + 1$, separated by a sheet of width a_i

$$\varepsilon_{i+1} - \varepsilon_i = \frac{1}{E}\left(\frac{F_{i+1}}{\bar{A}_{i+1}} - \frac{F_i}{\bar{A}_i}\right) = \frac{2 + v}{t_f E}\frac{dq_i}{dx}\, a_i \qquad (6.18)$$

Considering the equilibrium of a differential length of a typical bar i, an expression relating the axial force in the bar and the shear flows in the adjacent sheet segments is obtained.

$$\frac{dF_i}{dx} + q_i - q_{i-1} = 0 \qquad (6.19)$$

Now harmonic analysis can be applied directly to the case of simply-supported beams and cantilevers may be analysed by first establishing a substitute beam (Křístek, 1979b; Křístek and Evans, 1983). This and the analysis of other structural systems (e.g. continuous beams) is discussed in Křístek and Evans (1983).

In the harmonic analysis, the desired functions are anticipated in the form of a Fourier series with unknown amplitudes:

bar forces

$$F_i = \sum_{j=1}^{\infty} S_{i,j} \sin\frac{j\pi x}{L} \qquad (6.20)$$

sheet shear flows

$$q_i = \sum_{j=1}^{\infty} Q_{i,j} \cos\frac{j\pi x}{L} \qquad (6.21)$$

Since the external load factors can also be expressed by a Fourier series, the entire analysis can be conducted for each term of the series independently and the results simply added together.

Equations (6.18) and (6.19) may be written for any term of the series in the form of the following chain formulae:

$$S_{i+1,j} = \bar{A}_{i+1} \left(\frac{S_{i,j}}{\bar{A}_i} - \frac{2+v}{t_f} a_i \alpha_j Q_{i,j} \right) \tag{6.22}$$

and

$$Q_{i,j} = Q_{i-1,j} - \alpha_j S_{i,j} \tag{6.23}$$

where $\alpha = j\pi/L$.

The solution depends upon the arrangement of the stiffeners: several different arrangements are possible.

If a stiffener is situated at the mid-point of the flange width, because of the transverse symmetry, the shear flow in the two sheet segments on either side of the middle stiffener must be equal and, such that from eqn (6.23)

$$Q_{1,j} = -\frac{\alpha_j}{2} S_{1,j} \tag{6.24}$$

i.e. the unknown amplitude of the shear flow in the first sheet segment has been expressed in terms of the unknown amplitude of the axial force in the central bar. By substituting from eqn (6.24) into eqn (6.22) the unknown amplitude of the axial force in the next bar may be expressed again in terms of the unknown force in the central bar.

The procedure may be repeated and equations similar to eqns (6.22) and (6.23) established to relate the shear flows in all sheet segments and the axial forces in all bars to the first unknown force $S_{1,j}$. The value of $S_{1,j}$ is determined from the last equation established, i.e. when the edge of the flange is reached

$$Q_{r,j} = Q_{r-1,j} - \alpha_j S_{r,j} = Q_{e,j} \tag{6.25}$$

This equates the shear flow at the outside of the last (r-th) bar with the known amplitude of shear flow $Q_{e,j}$ acting between the edge of the flange and the top of the web (see eqn (6.12)).

Once the values of the forces in each of the bars are known, the stresses are calculated directly from $\sigma_{i,j} = S_{i,j}/\bar{A}_i$ and the shear flows in the sheet segments also may be calculated if required.

For the different arrangements of stiffeners where there is no central bar the condition at the mid-point of the flange is different. In such a case, there are no shear stresses in the central segment because of symmetry. Therefore, $Q_{0,j} = 0$ and from eqn (6.23) $Q_{1,j} = -\alpha_j S_{1,j}$; the remaining procedure is the same as in the preceding case.

The method described above can be applied by direct numerical evaluation using any pocket calculator without the need to solve any system of equations.

6.2.7 Some Other Methods of Shear Lag Analysis
Several methods have been proposed in which the stiffened flange plates are analysed as equivalent orthotropic plates. Such methods were improved by Abdel-Sayed (1969) who allowed for the true influence of the orthotropic stiffening. A similar procedure that considered a complete box girder was proposed by Malcolm and Redwood (1970). These methods also employ Fourier series analysis.

6.3 INFLUENCE OF VARIOUS PARAMETERS

6.3.1 Variation of Width/Span Ratio
It is well-known that as the flange width increases in relation to the span the shear lag effect becomes more pronounced.

6.3.2 The Type and Position of Loading
The non-uniformity of distribution of longitudinal stresses increases rapidly in the region of a point load or a support. Moving the load system away from the mid-span results in the reduction of the effective breadth ratios (Moffat and Dowling, 1975). The effective breadth ratios are only sensitive to the loaded length if this length is less than half of the span (Moffat and Dowling, 1975).

6.3.3 Effect of Stiffeners
Because stiffeners contribute to the axial load-carrying capacity of a flange without increasing its shear capacity, shear lag is more pronounced in a stiffened flange than in a flange without stiffening. It has been found (Moffat and Dowling, 1975; Křístek and Evans, 1983) that the effective breadth ratios decrease by a significant amount (in practically linear manner) as the stiffening factor increases.

For practical purposes it is desirable to simplify the numerical analysis by smearing the stiffener properties over the flange width. This approach must be verified from the point of view of:

(i)   the number of longitudinal ribs;
(ii) the effect of their own flexural rigidities and the eccentricity of their connections to the flange sheet;
(iii) the regularity of stiffener spacing.

It has been found that the concept of smearing the stiffener properties is acceptable for not only closely-spaced stiffeners but also, with little loss of accuracy for large, rather widely-spaced stiffeners. The conditions for the acceptability of such an approach are the regularity of the stiffener arrangement (Evans and Křistek, 1983) and the assumption that the stiffeners are concentrated in the flange plane (Fig. 6.8(a)).

Longitudinal stiffeners are generally welded to the inner side of the flange sheet (Fig. 6.8(b)). Due to the eccentricity of the stiffener connection individual portions of the flange with stiffeners which are eccentrically affected by shear flows acting in the plane of the flange sheet (Fig. 6.8(c)) tend to exhibit additional flexure, as shown in Fig. 6.9(d).

The influence of eccentricity of the stiffener connection is illustrated in Fig. 6.9. A steel box girder without intermediate diaphragms, under uniform loading ($w = 1\,\mathrm{N/mm}$), span $L = 9144\,\mathrm{mm}$ and with the cross-section shown in Fig. 6.9(a) is studied. The distribution of longitudinal stresses across the flange width is shown by the solid line in Fig. 6.9(b); the dashed line corresponds to that of the stiffeners concentrated in the flange sheet. Different flexural actions of individual stiffeners with adjacent flange portions are shown in Fig. 6.9(c) which depicts distributions of the longitudinal stresses along the stiffener depths.

It can be seen from Fig. 6.9(c) that the eccentrically-connected stiffeners, particularly those near the mid-point of the flange, exhibit stress distribution tending to that of a beam stressed by bending. The eccentricity

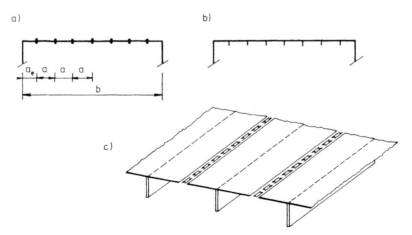

FIG. 6.8.   Stiffeners concentrated at the flange sheet and eccentrically-connected stiffeners.

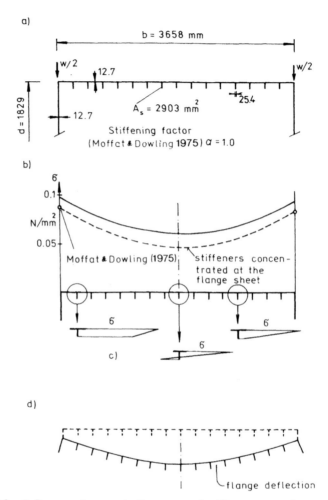

FIG. 6.9.   Influence of eccentrically-connected stiffeners upon the stress distribution and deflections.

of the stiffener connections thus results in the loss of efficiency of the total stiffened flange. This is also the reason why the solid line in Fig. 6.9(b), indicating the stress distribution for the eccentrically-connected stiffeners, falls completely (i.e. along the total width of the flange) above the dashed curve which corresponds to the fully-acting flange with stiffeners concentrated at the flange sheet.

The transverse flexure of the stiffened flange due to stiffener eccentricity is shown in Fig. 6.9(d). This kind of additional deformation is only partially restrained by a rather flexible flange.

It is seen that the stiffener eccentricity influences adversely the distribution of the longitudinal stresses unless closely-spaced sufficiently rigid transverse diaphragms are used to ensure equal deflection of all stiffeners. Thus the diaphragms influence indirectly the shear lag effects. Their presence is essential to enable one to use the methods that do not regard the stiffener eccentricity.

It has been found by Evans and Křístek (1983) that regularity of stiffener spacings (even if they provide the same total cross-sectional area of stiffeners and thus the same total second moment of area of the whole cross-section) influence the shear lag behaviour of the stiffened plate.

Figure 6.10 shows the cross-section of two steel box girders having a span $L = 9144$ mm, with stiffeners at different positions, loaded by uniformly-distributed loading of an intensity $w = 1$ N/mm. Although the longitudinal stress on the edge of the flange at mid-span for the case shown in Fig. 6.10(a) is $-0.087\,55$ N/mm^2, the stress for the stiffener arrangement shown in Fig. 6.10(b) reaches a magnitude of -0.0818 N/mm^2 only (see Evans and Křístek, 1983). According to Moffat and Dowling (1975) where a distinction is not made between the stiffener arrangements, for the stress effective breadth ratio 0.67 the corresponding stress is -0.082 N/mm^2. This represents an excellent agreement with the results obtained for the case shown in Fig. 6.10(b). Here, the stiffeners are situated at mid-points of adjacent flange portions and it is in accordance with a regular stiffener system assumed in Moffat and Dowling (1975).

However, the stiffener arrangement shown in Fig. 6.10(a) (with the same

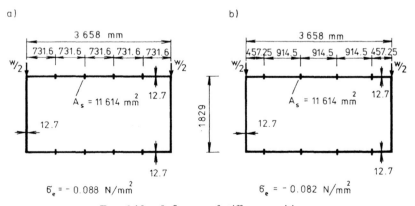

FIG. 6.10.   Influence of stiffener positions.

distances between all stiffeners and between the first stiffener and the web) where the stiffeners are situated more at the middle region of the flange, results in a 7 % increase in values of the longitudinal stresses compared with the case shown in Fig. 6.10(b). The reason for this is that the shear lag effect depends on shear deformability of those flange segments where the shear stress is of highest intensity, i.e. in the regions close to the webs. The width of the flange segment between the web and the first stiffener (and thus its shear deformability) is considerably lower in the case shown in Fig. 6.10(b) than that shown in Fig. 6.10(a).

The results clearly confirm the necessity to account for the actual stiffener positions for flanges with large, wide-spaced stiffeners. This fact cannot be accounted for by any method of analysis which assumes a regularly arranged structure, even if the Finite element method is used.

6.3.4 Initial Imperfections

It has been found by Křístek and Škaloud (1983) that in box girders without closely-spaced sufficiently rigid transverse diaphragms the distribution of longitudinal stresses across the width of the flange may be considerably influenced by initial imperfections of the flange sheet (Figs. 6.11(a) and (b)). This effect is not only due to a change in the second moment of area of the overall cross-section with imperfections and varying distance of a point at the flange from the cross-sectional neutral axis but due also to the shell action of the non-planar flange sheet loaded along its edges (where the flange is interconnected to the webs) by longitudinal shear flows (Fig. 6.11(d)).

An estimation of the influence of imperfections may be determined from Fig. 6.12 which shows a comparison between the distribution of the longitudinal stresses at mid-span of the considered box girder having the flange with two kinds of imperfection (as shown in Figs. 6.11(a) and (b)) of the same severity $\delta = \beta/100$. It can be seen that the shape of the imperfect surface of the flange affects the stress distribution pattern more than the absolute severity of the imperfection δ itself. Thus, imperfections, as shown in Fig. 6.11(b) (although δ is the same), have more adverse effects than those shown in Fig. 6.11(a). This is in agreement with the well-known fact that a decrease in the shear load-carrying capacity occurs in a corrugated sheet (Fig. 6.11(c)) whose axial load capacity remains unchanged and results in more pronounced shear lag effects than that of perfectly-plane flanges (similar to a flange orthotropy).

Here, it should be emphasised again, the important favourable influence of transverse diaphragms. These prevent the additional changes in the

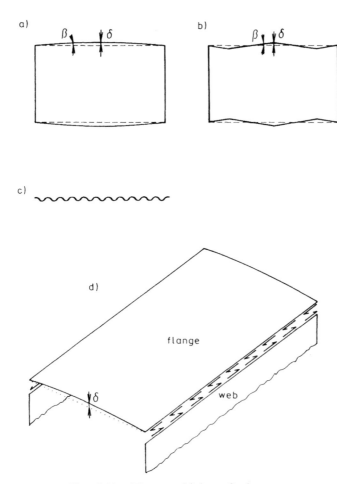

FIG. 6.11.   Flanges with imperfections.

flange shape and eliminate the shell effect of the flange indicated in Fig. 6.11(d). This effect is similar to that of flanges with eccentrically-connected stiffeners.

6.3.5 Influence of Loads Acting above Longitudinal Stiffeners

When calculating shear lag effects it is usually assumed that the load acts directly above webs. It requires a perfect transfer of all loads in the transverse direction (usually through a system of transverse diaphragms). However, cases may be encountered in design practice where short-span

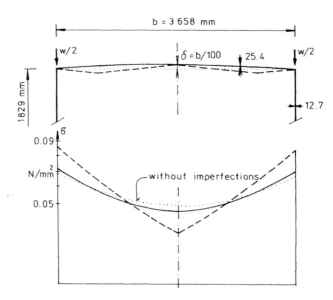

Fig. 6.12.   Influence of various kinds of flange imperfection upon the distribution
of longitudinal stresses.

bridges provided with rather stiff longitudinal ribs (e.g. railway bridges) are
loaded directly above these stiffeners.

An example of such an arrangement is shown in Fig. 6.13(b). Here, a
uniformly-loaded girder with central transverse diaphragm is considered.
The distribution of longitudinal stresses at the quarter-span for the load
acting above stiffeners (Fig. 6.13(b)) is compared in Fig. 6.13(c) with the
common case when the load acts above webs (Fig. 6.13(a)). An entirely
different character of stress-distribution pattern is apparent. The directly-
loaded stiffeners behave nearly as independent girders with typical stress
distribution along their depths accompanied by their own shear lag effects.

6.3.6 Influence of Overhanging Flanges

It is common practice in the design of steel box girder bridges that the webs
and flanges are interconnected as shown in Fig. 6.14(c). The influence of
this arrangement upon the distribution of longitudinal stresses in
comparison with the distribution corresponding to the case usually
considered (Fig. 6.14(b)) is shown in Figs. 6.14(d) and (e). As can be
expected the interconnection shown in Fig. 6.14(c) exhibits a favourable

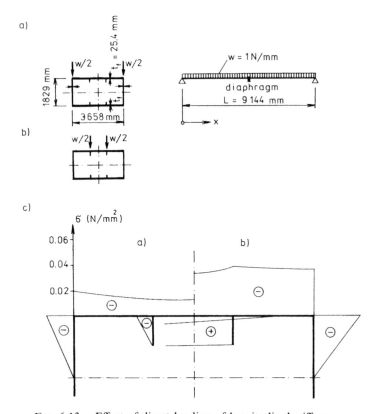

FIG. 6.13. Effect of direct loading of longitudinal stiffeners.

influence upon the distribution of longitudinal stress (in this case a decrease in the peak value by 9 %).

6.3.7 Curved Box Girders

The non-uniform character of the longitudinal stress distribution is similar to that of straight girders. Figure 6.15 gives a comparison of the results studying both the behaviour of a curved and a straight box girder (Křístek et al., 1981). The uniformly-distributed load is placed above the web segments. Although the intensity of loads acting on the individual webs is different, the resultant load on each web is the same. As can be seen these stresses tend to increase nearer the inner web. This effect is due to a pronounced torsion in the girder, different lengths and is due consequently to different rigidities of the outer and inner webs.

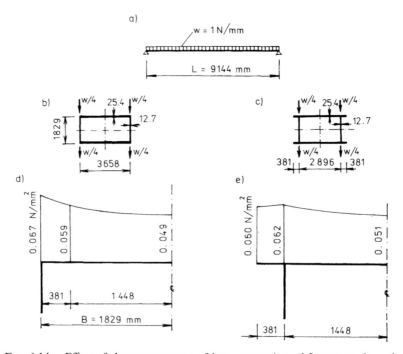

FIG. 6.14.   Effect of the arrangement of interconnection of flanges to the webs.

6.3.8. Method of Reducing Shear Lag Effects

Shear lag essentially involves a loss of efficiency of the material used in the flange. Hence, it is desirable to look for methods to reduce its effect.

Shear lag is a result of shear deformations of the flange sheet that depends upon the shear stress intensity. The shear stresses can be reduced by enlarging the flange thickness in the regions adjacent to the webs. In doing so the total amount of material used on the flange can be sometimes kept constant. It is sufficient to place more material in boundary zones of the flange, as shown in Fig. 6.16(b) (compare with the original version in Fig. 6.16(a)). This results in more uniform distribution of the longitudinal stresses as shown in Fig. 6.16(c)—the left upper-quarter of the cross-section (see Curve b) compared to the original distribution (Curve a). The corresponding increase in the load-carrying capacity of the girder is 14% (Křístek *et al.*, 1981).

In the case of a steel girder the variation of the flange thickness t_f can be achieved usually in steps (Fig. 6.16(d)). If this variation is idealised by an

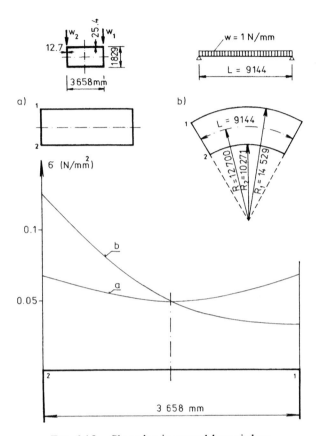

FIG. 6.15. Shear lag in curved box girders.

equivalent parabolic variation of degree n the distributions of longitudinal stresses as in Fig. 6.16(e) are obtained. It has been found by Škaloud and Křístek (1981) that a suitable variation of the flange thickness (a variation corresponding to the parabola of the second or third-degree seems to be an optimum) results in a reduction of the maximum longitudinal stress at the flange edge as well as in an increase in stresses in the middle region of the flange. This means that a more uniform and consequently more favourable distribution of the longitudinal stresses across the flange width is achieved.

Since the increase in the thickness of the flange near the edges can be both costly and needs a lot of welding it is easier—in the case of a flange with stiffeners—to place the last stiffeners at a smaller distance away from the web (see also Fig. 6.10(b)).

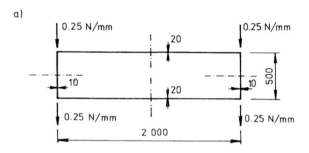

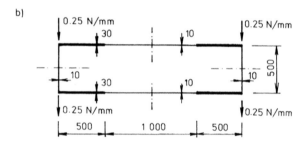

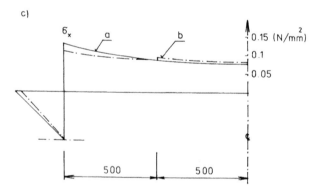

FIG. 6.16.   A possible reduction of the effect of shear lag by means of a suitable variation of the flange thickness.

d)

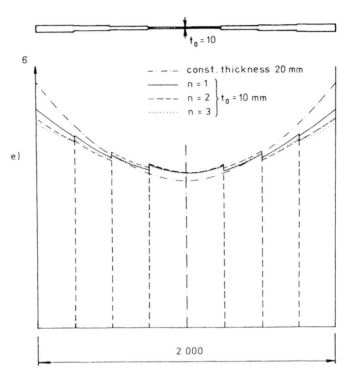

FIG. 6.16—contd.

The conclusions given above are valid for flanges in tension. If a flange is subjected to compression the weakened central part of the flange is more liable to buckle; therefore the whole problem ought to be solved with due regard to the interaction of shear lag and plate buckling and this is studied fully in Škaloud and Křistek (1981).

6.3.9 Some Other Aspects of Shear Lag

Conditions in the structure are changed if higher stress levels beyond the elastic range are reached. The elastic–plastic effects are regarded in individual Codes to a different degree. Recent studies (Professor Dowling, Dr Lamas) have found that for very short, wide boxes the shear capacity of the flange could be exhausted before the reserve strength of the centre of that flange had been mobilised. For longer boxes, plasticity develops until finally the full width of the flange is affected.

The added complexities due to effects of large deflections and the interaction of shear lag and plate buckling (Škaloud and Křístek, 1981) are not discussed here as this study is intended for use by practising engineers and is meant as general information regarding the shear lag phenomenon itself.

6.4 CONCLUSIONS

1.   Some suitable methods for the shear lag analysis have been reviewed.

   (a)   It has been shown that the Finite element method may be applied but it requires the use of large computers and is very expensive in terms of computer time and storage.

   (b)   For girders of constant cross-section the Folded plate theory which takes advantage of harmonic analysis may be applied to a variety of support conditions; a considerable reduction of computer time results.

   (c)   Methods that enable shear lag effects to be predicted from hand calculations are suitable for use during the design stage when several girders may be analysed to determine the optimum proportions.

2.   It follows from the results of studying the influence of various parameters that the arrangement of longitudinal stiffeners may considerably affect the distribution of longitudinal stresses across the flange width.

   (a)   It is reasonable to assume the stiffener properties to be distributed over the flange width only if the stiffeners are situated regularly at the flange sheet and in the case of eccentrically-connected longitudinal stiffeners if a system of sufficiently rigid transverse stiffeners is used in order to eliminate additional transverse flexure of the flange caused by the stiffener eccentricity.

   (b)   Similar effects are encountered when analysing initial imperfections.

3.   This study has concentrated mainly on the distribution of longitudinal stresses across the flange width. However, the shear lag manifests itself also in several further effects, e.g. in the

additional flexibility of the girder which results in an increase of deflections and which alters the overall bending moment and shear force diagrams in statically-indeterminate structures. The shear lag also affects the pattern of distribution of shear stresses in the flange.

(a)   For girders of practical proportions the changes of the overall bending moment diagrams in statically-indeterminate girders is usually insignificant and may be neglected in approximate methods intended as design tools.

(b)   The influence of shear lag upon the increase of deflections is however of considerable importance. The general methods of analysis can also be used to provide an adequate picture of deformations of the structure. The harmonic series type solutions take advantage of an extremely rapid convergence in deflections. When adopting the effective breadth concept the design rules based on finite element results obtained by Moffat and Dowling (1975) may be used.

REFERENCES

ABDEL-SAYED, G. (1969) Effective width of steel deck-plate in bridges. *Journal of the Structural Division, ASCE*, **95**, ST7.

DEFRIES-SKENE, A. and SCORDELIS, A. C. (1964) Direct stiffness solution for folded plates. *Journal of the Structural Division, ASCE*, **90**, ST4.

DEPARTMENT OF THE ENVIRONMENT (1973) *Inquiry into the Basis of Design and Method of Erection of Steel Box Girder Bridges*, HMSO, London.

EVANS, H. R. and KŘÍSTEK, V. (1983) A hand calculation of the shear lag effect in flanges with large, widely-spaced stiffeners. *Proc. Instn Civ. Engrs*, to be published.

EVANS, H. R. and TAHERIAN, A. R. (1977) The prediction of the shear lag effect in box girders. *Proc. Instn Civ. Engrs*, Part 2, 63.

EVANS, H. R. and TAHERIAN, A. R. (1980) A design aid for shear lag calculations. *Proc. Instn Civ. Engrs*, Part 2, 69.

GIRKMANN, K. (1954) *Flächentragwerke*, Springer-Verlag, Wien.

KŘÍSTEK, V. (1979a) *Theory of Box Girders*, J. Wiley and Sons, New York.

KŘÍSTEK, V. (1979b) Folded plate approach to analysis of shear wall systems and frame structures. *Proc. Instn Civ. Engrs*, Part 2, 67.

KŘÍSTEK, V. and EVANS, H. R. (1983) A hand calculation of the shear lag effect in unstiffened flanges and in flanges with closely spaced stiffeners. *Civil Engineering for Practicing and Design Engineers*, in press.

KŘÍSTEK, V. and ŠKALOUD, M. (1983) Effect of an initial curvature on the shear lag phenomenon in wide flanges. *Acta Technica ČSAV*, 1, in press.

KŘÍSTEK, V., STUDNIČKA, J. and ŠKALOUD, M. (1981) Shear lag in wide flanges of steel bridges. *Acta Technica ČSAV*, Prague, No. 4.

KUHN, P. (1956) *Stresses in Aircraft and Shell Structures*, McGraw-Hill, New York.

LIN, C. S. and SCORDELIS, A. C. (1971) Computer program for bridges on flexible bents. Report No. 71-24, UC-SESM, University of California, Berkeley.

MALCOLM, D. J. and REDWOOD, R. G. (1970) Shear lag in stiffened box girders. *Journal of the Structural Division, ASCE*, **96**, ST7.

MEYER, C. and SCORDELIS, A. C. (1970) Analysis of curved folded plate structures. Report No. 70-8, UC-SESM, University of California, Berkeley.

MOFFAT, K. R. and DOWLING, P. J. (1975) Shear lag in steel box girder bridges. *Structural Engineer*, **53**(10), 439–48.

ŠKALOUD, M. and KŘÍSTEK, V. (1981) Stability problems of steel box girder bridges. *Academia, Prague*.

TAHERIAN, A. R. and EVANS, H. R. (1977) The bar simulation method for the calculation of shear lag in multi-cell and continuous box girders. *Proc. Instn Civ. Engrs*, Part 2, 63.

Chapter 7

COMPRESSIVE STRENGTH OF BIAXIALLY LOADED PLATES

R. Narayanan

*Department of Civil and Structural Engineering,
University College, Cardiff, UK*

and

N. E. Shanmugam

*Department of Civil Engineering,
National University of Singapore, Singapore*

SUMMARY

An approximate method to predict the post-buckling and collapse behaviour of plates loaded axially along two mutually perpendicular directions is presented. The theory is based on the Energy method and accounts for the initial imperfection, plate slenderness and varying rates of strain in the two directions. Interaction curves for square plates subjected to longitudinal and transverse stresses, applicable to six combinations of boundary conditions, have been presented. The results have been compared with more exact methods and are found to be of acceptable accuracy, for practical structures having small imperfections.

NOTATION

A A coefficient governing the deflection amplitude
A_0 A coefficient governing the initial deflection amplitude
a, b Length and width of the plate in the x and y directions

D $Et^3/12(1 - v^2)$

E Modulus of elasticity of steel

$f(y)$ A function of y defining the deflected shape

$g(x)$ A function of x defining the deflected shape

m Magnification ratio A/A_0

p, q Number of half-wave lengths in the x and y directions

t Thickness of the plate

U_b Strain energy due to bending

U_s Strain energy due to strain in the middle plane

U_{int} Internal strain energy $(U_b + U_s)$

w Deflection at any point (x, y) of the plate

w_0 Initial imperfection at any point (x, y) of the plate

v Poisson's ratio

δ_0 Maximum value of initial plate imperfection

α b/a

ε_x Strain in the x-direction

ε_y Strain in the y-direction

σ_e $\pi^2 E/12(1 - v^2)$

σ_{eq} Equivalent stress

σ_x Longitudinal stress given by $\sigma_{xa} + \sigma_{xy}$

σ_{xa} Applied stress in the x-direction

σ_{xy} Induced stress in the x-direction due to the action of σ_{ya}

σ_y Transverse stress given by $\sigma_{ya} + \sigma_{yx}$

σ_{ya} Applied stress in the y-direction

σ_{yx} Induced stress in the y-direction due to the action of σ_{xa}

σ_{ys} Yield stress of steel

7.1 INTRODUCTION

It is well known that thin plates remain in stable equilibrium under loads greater than their buckling loads. Methods of determining the elastic critical loads for perfect plates, loaded in plane, have been available for many years (see Timoshenko, 1936). However, satisfactory methods of solving the *elastic* large deflection plate equations, including the effect of geometric imperfections, have been suggested only recently by Williams (1971, 1975). Analytical techniques to study the *elastic–plastic* behaviour of plates have also been developed recently by Zienkiewicz *et al.* (1969), Crisfield (1973), Yam (1974) and Dowling (1974). These methods are useful for the design of plates loaded longitudinally in plane and carrying no transverse in-plane loads and no normal loads. The methods developed so far are extravagant in computer time and, are, therefore expensive where a number of parametric studies are required to be carried out in the early stages of design. An approximate method of evaluating the strength of a plate under biaxial compression is presented in this chapter and is suitable for any combination of boundary conditions.

The method is considerably economical in terms of computing time and yet estimates the plate strengths without any significant loss in accuracy.

There are a number of structures such as double bottoms of ship hulls, dock gates and box girders where plate elements are subjected to biaxial loading, i.e. loads act both longitudinally and transversely in plane. A typical double-bottomed ship hull consists of inner and outer plating connected by thin webs which run longitudinally and transversely. Transverse webs (or 'floors') are typically spaced at 3 m centres, while longitudinal webs (or 'intercostals') are spaced at around 5 m centres; the plates are stiffened longitudinally by 'brackets'. The bulkheads or transverse diaphragms made up of stiffened plating divide the hull up longitudinally. The longitudinal compressive stresses into the hull plates are introduced due to the longitudinal hogging of the entire ship, when it passes over the peak of a wave. Transverse stresses are a consequence of the bending of the hull, under the action of hydrostatic pressure from below.

A similar state of stress occurs in certain box-girder flange plates. These plate panels carry longitudinal in-plane compression caused by the sagging moment in the box girder and transverse in-plane compression consequent on these panels acting as the compression flanges of the cross girder.

Smith (1975) has confirmed recently that little information on the behaviour of plates subjected to biaxial in-plane loading is available in published literature; the only systematic work done until his work appears to be a limited amount of experimental work and an approximate analytical method developed by Becker et $al.$ (1970).

As has already been stated, this chapter is mainly concerned with the biaxial compressive strength of plating used in the hulls of ships and (in certain instances) in box girder construction. The parameters studied include plate aspect ratio (a/b), initial imperfection of the plate, ratio of strain increment in the x and y directions $(\varepsilon_x/\varepsilon_y)$ and the plate slenderness (b/t). Plates having six combinations of boundary conditions (given in Table 7.1) are considered.

The approximate analysis developed in this paper is based on the Energy method and is a natural extension of the approximate methods developed earlier for uniaxial loading (see Horne and Narayanan, 1976; Narayanan and Shanmugam, 1980).

TABLE 7.1
BOUNDARY CONDITIONS OF VARIOUS CASES

Case	Edge condition	A_0
1	Simply supported at $x = 0$, $x = a$, and simply supported at $y = 0$, $y = b$	δ_0
2	Clamped at $x = 0$, $x = a$, and clamped at $y = 0$, $y = b$	$0.25\delta_0$
3	Simply supported at $x = 0$, $x = a$, and clamped at $y = 0$, $y = b$	$0.5\delta_0$
4	Simply supported at $x = 0$, clamped at $x = a$, and simply supported at $y = 0$, clamped at $y = b$	$14.8\delta_0$
5	Clamped at $x = 0$, $x = a$, simply supported at $y = 0$, and clamped at $y = b$	$1.92\delta_0$
6	Simply supported at $x = 0$, $x = a$ and $y = 0$, and clamped at $y = b$	$3.85\delta_0$

7.2   ASSUMPTIONS

The plates considered in this chapter are of proportions typically used in civil engineering and marine structures and have small initial imperfections. This would mean, for example, that plates supported on all the four edges do not have b/t values in excess of 70; the computations have, however, been continued for b/t values of up to 100.

In carrying out the analysis, the following assumptions are made:

1. The material of the plate is homogeneous, isotropic, elastic and thereafter perfectly plastic. The effect of strain hardening is neglected.
2. All the edges of the plates are taken to be held straight in plane and out-of-plane but are free from restraining or applied moments.
3. Membrane shear stresses on planes parallel to the edges of the plate panel are assumed to be zero.
4. The number and length of half-waves of buckling is assumed to be the same in post-buckled stages as at incipient buckling.
5. The second order membrane strains are assumed to be entirely dependent on the out-of-plane displacements. Although this assumption violates the membrane equilibrium and compatibility conditions locally, the integrated effects of this violation would be negligible; the results obtained from this analysis are found to be satisfactory.

6.   The values of displacements in the x and y directions are very small in comparison with transverse deflections in the z-direction and can be neglected when calculating the energy of the system.

7.3   ANALYSIS

In practice all plates have imperfections; the analytical treatment, therefore, deals with an initially imperfect plate. The actual imperfections in a plate, however, follow no general pattern; their effects are most detrimental to the load-carrying capacity of the plate if their pattern is similar in form to the deflected shape due to plate buckling. In view of this, it is usual to base the analysis on an initial shape similar in form to the buckled shape of the plate, so that a sensible and conservative strength prediction of the plate results.

Let us assume that the plate in Fig. 7.1 has an initial imperfection shape given by

$$w_0 = A_0 f(y) g(x) \qquad (7.1)$$

where A_0 is a coefficient defining the amplitude of the initial imperfection. (A_0 can be related to the maximum value of initial imperfection δ_0, by a constant, which can be obtained by substitution.) $f(y)$ and $g(x)$ are deflection functions in terms of y and x respectively and may be chosen in the form of infinite series or trigonometric or algebraic functions to suit the conditions of support along the longitudinal and transverse edges. A_0, the initial imperfection coefficient, can be related to the maximum value of

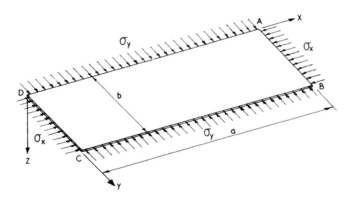

FIG. 7.1.   Biaxially loaded plate.

the initial imperfection, δ_0, in each of the cases considered by satisfying the boundary conditions of the plates. As an example, considering the case of an approximately square plate simply supported along all the four edges, w_0, the function describing the initial shape can be obtained by choosing

$$g(x) = \sin \frac{\pi x}{a} \quad \text{and} \quad f(y) = \sin \frac{\pi y}{b}$$

When $x = a/2$ and $y = b/2$, the maximum value of initial imperfection, δ_0 is obtained. In this case $\delta_0 = A_0$. Values of A_0 in terms of δ_0 for the various boundary conditions considered in this chapter are given in Table 7.1. Due to the action of loads, the deflected shape of the plate is assumed to be described by the equation

$$w = Af(y)g(x) \tag{7.2}$$

The magnification ratio, m, is therefore

$$m = \frac{A}{A_0}$$

Let us first consider a plate simply supported along all four edges with the edges held straight. The buckled shape is assumed to consist of a series of wave-forms in a chequer-board pattern.

Suitable functions which satisfy the loading edge condition are:

$$g(x) = \sin \frac{p\pi x}{a} \tag{7.3}$$

and

$$f(y) = \sin \frac{q\pi y}{b} \tag{7.4}$$

where p and q are the number of half-wave lengths in the x and y directions; a and b are the length and width of the plate respectively.

Considering an imperfect plate, after buckling has occurred (see Fig. 7.1), the longitudinal stress σ_x will vary across the width of the plate and the Poisson expansion in the y-direction is uniform along the edges parallel to the y-axis and of value $(v/E)\int_0^b \sigma_x \, dy$. In addition to this uniform expansion, there will be a uniform contraction e_{yx} given by:

$$e_{yx} = \frac{\sigma_{yx} b}{E} + (q\pi)^2 \frac{(A^2 - A_0^2)}{4b} \sin^2 \frac{p\pi x}{a} \tag{7.5}$$

where σ_{yx} is the induced stress in the y-direction along DA and BC. The second term in the right-hand side of eqn (7.5) is due to the flexural shortening.

For the boundary condition in which the edges are taken to be held straight:

$$\int_0^a \sigma_{yx}\, dx = 0 \qquad (7.6)$$

Integrating eqn (7.5) from 0 to a and using eqn (7.6):

$$e_{yx} = \frac{(q\pi)^2}{8b}(A^2 - A_0^2) \qquad (7.7)$$

Hence from eqn (7.5):

$$\sigma_{yx} = \frac{(q\pi)^2 E(A^2 - A_0^2)}{8b^2} \cos \frac{2p\pi x}{a} \qquad (7.8)$$

Similarly

$$\sigma_{xy} = \frac{(p\pi)^2 E(A^2 - A_0^2)}{8a^2} \cos \frac{2q\pi y}{b} \qquad (7.9)$$

The total stress acting in the x-direction is the summation of the stress in the x-direction due to the applied strain in the x-direction and the induced stress due to loading in the y-direction.

$$\sigma_x = \sigma_{xa} + \sigma_{xy}$$
$$= \sigma_{xa} + \frac{(p\pi)^2 E(A^2 - A_0^2)}{8a^2} \cos \frac{2q\pi y}{b} \qquad (7.10)$$

Similarly

$$\sigma_y = \sigma_{ya} + \frac{(q\pi)^2 E(A^2 - A_0^2)}{8b^2} \cos \frac{2p\pi x}{a} \qquad (7.11)$$

Considering a rectangular plate subjected in the post-buckling state to biaxial stresses σ_x and σ_y, which vary across the loading edges, the total strain energy (U_{int}) is the summation of energy due to bending and the energy due to strain in the middle plane of the plate:

$$U_{int} = U_b + U_s \qquad (7.12)$$

where U_b is the energy due to bending and U_s is the energy due to strain in the middle plane of the plate.

Considering a perfect plate, the bending energy stored during elastic critical buckling is approximately given by:

$$U_b = \frac{D}{2} \int_0^b \int_0^a \left(\frac{\partial^2 w}{\partial x^2} + \frac{\partial^2 w}{\partial y^2}\right)^2 dx\,dy$$

Substituting for w from eqn (7.2) we get:

$$U_b = \frac{\pi^4 A^2}{8} Dab \left(\frac{p^2}{a^2} + \frac{q^2}{b^2}\right)^2 \tag{7.13}$$

This solution has been obtained on the assumption that the plate is perfectly flat.

The strain energy stored in the imperfect plate due to bending is likewise obtained from the result of the perfect plate with A_0 included. It can be shown that

$$U_b = \frac{\pi^4 (A - A_0)^2 D}{8} ab \left(\frac{p^2}{a^2} + \frac{q^2}{b^2}\right)^2 \tag{7.14}$$

The energy U_s due to strain in the middle plane of the plate is given by

$$U_s = \frac{1}{2E} \int_0^b \int_0^a (\sigma_x^2 + \sigma_y^2 - 2v\sigma_x\sigma_y)t\,dx\,dy$$

$$= \frac{t}{2E} \left\{ a \int_0^b \sigma_{xa}^2\,dy + \frac{(p\pi)^2 E(A^2 - A_0^2)}{4a} \int_0^b \sigma_{xa}\cos\frac{2q\pi y}{b}\,dy \right.$$

$$+ \frac{(p\pi)^4 E^2(A^2 - A_0^2)^2 b}{128a^3} + b \int_0^a \sigma_{ya}^2\,dx$$

$$+ \frac{(q\pi)^2 E(A^2 - A_0^2)}{4b} \int_0^a \sigma_{ya}\cos\frac{2p\pi x}{a}\,dx$$

$$\left. + \frac{(q\pi)^4 E^2(A^2 - A_0^2)^2 a}{128b^3} - 2v \int_0^b \int_0^a \sigma_{xa}\sigma_{ya}\,dx\,dy \right\} \tag{7.15}$$

The total strain energy is obtained by adding U_b and U_s. A change in the total strain energy $(\mathrm{d}\,U_{\mathrm{int}})$ due to small increments in the amplitude from A to $(A + \mathrm{d}A)$ is obtained as:

$$\mathrm{d}\,U_{\mathrm{int}} = \frac{\pi^4(A - A_0)D}{4}\,ab\left(\frac{p^2}{a^2} + \frac{q^2}{b^2}\right)^2 \mathrm{d}A + \frac{at}{E}\int_0^b \sigma_{xa}\,\mathrm{d}\sigma_{xa}\,\mathrm{d}y$$

$$+ \frac{bt}{E}\int_0^a \sigma_{ya}\,\mathrm{d}\sigma_{ya}\,\mathrm{d}x + \frac{(p\pi)^2 tA\,\mathrm{d}A}{4a}\int_0^b \sigma_{xa}\cos\frac{2q\pi y}{b}\,\mathrm{d}y$$

$$+ \frac{(p\pi)^2 t(A^2 - A_0^2)}{8a}\int_0^b \mathrm{d}\sigma_{xa}\cos\frac{2q\pi y}{b}\,\mathrm{d}y$$

$$+ \frac{(q\pi)^2 tA\,\mathrm{d}A}{4b}\int_0^a \sigma_{ya}\cos\frac{2p\pi x}{a}\,\mathrm{d}x$$

$$+ \frac{(q\pi)^2 t(A^2 - A_0^2)}{8b}\int_0^a \mathrm{d}\sigma_{ya}\cos\frac{2p\pi x}{a}\,\mathrm{d}x$$

$$+ \frac{\pi^4 Et(A^2 - A_0^2)A\,\mathrm{d}A}{64a^3 b^3}\left((pb)^4 + (qa)^4\right)$$

$$- \frac{vt}{E}\int_0^b\int_0^a \sigma_{ya}\,\mathrm{d}\sigma_{xa}\,\mathrm{d}x\,\mathrm{d}y - \frac{vt}{E}\int_0^b\int_0^a \sigma_{xa}\,\mathrm{d}\sigma_{ya}\,\mathrm{d}x\,\mathrm{d}y \qquad (7.16)$$

The applied strain in the x-direction may be expressed as the sum of change due to σ_x, the Poisson expansion and that due to flexural shortening:

$$\varepsilon_x = \frac{\sigma_x}{E} - \frac{v}{aE}\int_0^a \sigma_{ya}\,\mathrm{d}x + \frac{(p\pi)^2(A^2 - A_0^2)}{4a^2}\sin^2\frac{q\pi y}{b}$$

Substituting for σ_x from eqn (7.10):

$$\varepsilon_x = \frac{1}{E}\left\{\sigma_{xa} + \frac{(p\pi)^2 E(A^2 - A_0^2)}{8a^2}\cos\frac{2q\pi y}{b}\right\} - \frac{v}{aE}\int_0^a \sigma_{ya}\,\mathrm{d}x$$

$$+ \frac{(p\pi)^2(A^2 - A_0^2)}{4a^2}\sin^2\frac{q\pi y}{b} \qquad (7.17)$$

Hence, the change in the applied strain $\mathrm{d}\varepsilon_x$ is obtained from

$$\mathrm{d}\varepsilon_x = \frac{1}{E}\left\{\mathrm{d}\sigma_{xa} + \frac{(p\pi)^2 EA\,\mathrm{d}A}{4a^2}\cos\frac{2q\pi y}{b}\right\}$$

$$- \frac{v}{aE}\int_0^a \mathrm{d}\sigma_{ya}\,\mathrm{d}x + \frac{(p\pi)^2 A\,\mathrm{d}A}{2a^2}\sin^2\frac{q\pi y}{b}$$

This equation can be further simplified as

$$d\varepsilon_x = \frac{d\sigma_{xa}}{E} + \frac{(p\pi)^2 A \, dA}{4a^2} - \frac{v}{aE} \int_0^a d\sigma_{ya} \, dx \tag{7.18}$$

Similarly, the change in the applied strain $d\varepsilon_y$ is given by

$$d\varepsilon_y = \frac{d\sigma_{ya}}{E} + \frac{(q\pi)^2}{4b^2} A \, dA - \frac{v}{bE} \int_0^b d\sigma_{xa} \, dy \tag{7.19}$$

The total work done by the loads in both x and y directions due to small increment in strains in the respective directions is given by

$$dT = a \, d\varepsilon_x \int_0^b \sigma_x t \, dy + b \, d\varepsilon_y \int_0^a \sigma_y t \, dx$$

$$= \frac{at}{E} \int_0^b \sigma_{xa} d\sigma_{xa} \, dy + at \int_0^b \frac{(p\pi)^2 (A^2 - A_0^2)}{8a^2} d\sigma_{xa} \cos \frac{2q\pi y}{b} \, dy$$

$$+ at \int_0^b \frac{(p\pi)^2 A}{4a^2} \sigma_{xa} \, dA \, dy - \frac{vt}{E} \int_0^b \int_0^a \sigma_{xa} d\sigma_{ya} \, dx \, dy$$

$$+ \frac{bt}{E} \int_0^a \sigma_{ya} d\sigma_{ya} \, dx + bt \int_0^a \frac{(q\pi)^2 (A^2 - A_0^2)}{8b^2} d\sigma_{ya} \cos \frac{2p\pi x}{a} \, dx$$

$$+ bt \int_0^a \frac{(q\pi)^2 A}{4b^2} \sigma_{ya} \, dA \, dx - \frac{vt}{E} \int_0^a \int_0^b \sigma_{ya} d\sigma_{xa} \, dy \, dx \tag{7.20}$$

Using $d U_{int} = dT$, we obtain from eqns (7.16) and (7.20):

$$\frac{\pi^4 (A - A_0) D}{4} ab \left(\frac{p^2}{a^2} + \frac{q^2}{b^2} \right)^2 dA + \frac{\pi^4 Et (A^2 - A_0^2) A \, dA}{64 a^3 b^3} ((pb)^4 + (qa)^4)$$

$$= \frac{(p\pi)^2 tA \, dA}{4a} \int_0^b \sigma_{xa} \left(1 - \cos \frac{2q\pi y}{b} \right) dy$$

$$+ \frac{(q\pi)^2 tA \, dA}{4b} \int_0^a \sigma_{ya} \left(1 - \cos \frac{2p\pi x}{a} \right) dx \tag{7.21}$$

Putting

$$D = \frac{Et^3}{12(1 - v^2)} \qquad \alpha = \frac{b}{a}$$

$$m = \frac{A}{A_0} \qquad \sigma_e = \frac{\pi^2 E}{12(1 - v^2)}$$

eqn (7.21) can be rearranged to give

$$\frac{\sigma_e}{E}\left(\frac{t}{b}\right)^2 (m-1)\frac{(p^2\alpha^2+q^2)^2}{\alpha} + \frac{\pi^2}{16}\frac{A_0^2}{b^2}(m^3-m)\frac{(\alpha^4p^4+q^4)}{\alpha}$$

$$= \frac{2p^2m}{aE}\int_0^b \sigma_{xa}\sin^2\frac{q\pi y}{b}\,dy + \frac{2q^2m}{bE}\int_0^a \sigma_{ya}\sin^2\frac{p\pi x}{a}\,dx \qquad (7.22)$$

Considering the effect of an increment in amplitude by $dm = dA/A_0$:

$$dm\,\frac{(p^2\alpha^2+q^2)^2}{\alpha}\left(\frac{t}{b}\right)^2\frac{\sigma_e}{E} + \frac{\pi^2}{16}\frac{A_0^2}{b^2}\frac{(\alpha^4p^4+q^4)}{\alpha}(3m^2-1)\,dm$$

$$= \frac{2p^2}{aE}\,dm\int_0^b \sigma_{xa}\sin^2\frac{q\pi y}{b}\,dy + \frac{2p^2m}{aE}\int_0^b d\sigma_{xa}\sin^2\frac{q\pi y}{b}\,dy$$

$$+ \frac{2q^2}{bE}\,dm\int_0^a \sigma_{ya}\sin^2\frac{p\pi x}{a}\,dx + \frac{2q^2m}{bE}\int_0^a d\sigma_{ya}\sin^2\frac{p\pi x}{a}\,dx$$

$$(7.23)$$

Using $d\sigma_{xa}$ and $d\sigma_{ya}$ derived from eqns (7.18) and (7.19), eqn (7.23) becomes

$$\frac{dm(p^2\alpha^2+q^2)^2}{\alpha}\left(\frac{t}{b}\right)^2\frac{\sigma_e}{E} + \frac{\pi^2 A_0^2}{16b^2}\frac{(\alpha^4p^4+q^4)}{\alpha}(3m^2-1)\,dm$$

$$= \frac{2p^2}{aE}\,dm\int_0^b \sigma_{xa}\sin^2\frac{q\pi y}{b}\,dy + \frac{2q^2}{bE}\,dm\int_0^a \sigma_{ya}\sin^2\frac{p\pi x}{a}\,dx$$

$$+ p^2m\alpha\,d\varepsilon_x + \frac{q^2m}{\alpha}\,d\varepsilon_y - \frac{p^4\pi^2 A_0^2 m^2\alpha\,dm}{4a^2} - \frac{q^4\pi^2 A_0^2 m^2\,dm}{4\alpha b^2}$$

$$+ \frac{2p^2mv}{a^2E}\int_0^b\int_0^a d\sigma_{ya}\sin^2\frac{q\pi y}{b}\,dx\,dy + \frac{2q^2mv}{b^2E}\int_0^b\int_0^a d\sigma_{xa}\sin^2\frac{p\pi x}{a}\,dx\,dy$$

$$(7.24)$$

The first two integrals of the right-hand side of the equation can be eliminated between eqns (7.22) and (7.24) and after performing the integration and rearranging the terms, we obtain the following equation:

$$p^2\alpha\,d\varepsilon_x + \frac{q^2}{\alpha}\,d\varepsilon_y = \frac{(p^2\alpha^2+q^2)^2}{\alpha}\left(\frac{t}{b}\right)^2\frac{\sigma_e}{E}\frac{dm}{m^2} + \frac{3\pi^2}{8}\left(\frac{A_0}{b}\right)^2\frac{(\alpha^4p^4+q^4)}{\alpha}m\,dm$$

$$- \frac{v}{E}\left\{\frac{\alpha p^2}{a}\int_0^a d\sigma_{ya}\,dx + \frac{q^2}{\alpha b}\int_0^b d\sigma_{xa}\,dx\right\} \qquad (7.25)$$

The $d\sigma_{xa}$ corresponding to the applied strain $d\varepsilon_x$ can be obtained as

$$d\sigma_{xa} = E \left\{ d\varepsilon_x - \frac{(p\pi)^2 A_0^2 m \, dm}{4a^2} + \frac{v}{aE} \int_0^a d\sigma_{ya} \, dx \right\} \qquad (7.26)$$

Similarly, the applied stress σ_{ya} corresponding to the applied strain $d\varepsilon_y$ can be evaluated:

$$d\sigma_{ya} = E \left\{ d\varepsilon_y - \frac{(q\pi)^2 A_0^2 m \, dm}{4b^2} + \frac{v}{bE} \int_0^b d\sigma_{xa} \, dy \right\} \qquad (7.27)$$

By solving eqns (7.25), (7.26) and (7.27) numerically for the values of σ_{xa} and σ_{ya}, the stresses σ_x and σ_y at any section can be obtained from eqns (7.10) and (7.11). A number of special cases of the above theory will now be derived.

Case 1
As a very simple illustration of the use of the above equations, let us consider the case of a rectangular plate simply supported along all the four edges subjected to end long compression in the x-direction only. We set $d\varepsilon_y$ and $d\sigma_{ya}$ equal to zero and consider only a single half-wave buckled mode, i.e. $q = p = 1$. Equation (7.27) becomes:

$$\frac{v}{bE} \int_0^b d\sigma_{xa} \, dy = \frac{(q\pi)^2 A_0^2 m \, dm}{4b^2} \qquad (7.28)$$

Substituting eqn (7.28) into eqn (7.25):

$$\alpha \, d\varepsilon_x = \frac{(\alpha^2 + 1)^2}{\alpha} \left(\frac{t}{b}\right)^2 \frac{\sigma_e}{E} \frac{dm}{m^2} + \frac{\pi^2 A_0^2}{8b^2} m \, dm \left(3\alpha^3 + \frac{1}{\alpha}\right) \qquad (7.29)$$

Integrating eqn (7.29) and replacing σ_e by $\pi^2 E/[12(1 - v^2)]$ and α by b/a:

$$\varepsilon_x = \frac{(m-1)}{m} \left(\frac{t}{b}\right)^2 \frac{\pi^2}{12(1 - v^2)} \left(\frac{b^2}{a^2} + \frac{a^2}{b^2} + 2\right) + \frac{\pi^2 A_0^2}{16a^2} \left(3 + \frac{a^4}{b^4}\right) (m^2 - 1) \qquad (7.30)$$

and

$$\sigma_x = E \left[\varepsilon_x - (m^2 - 1) \frac{\pi^2 A_0^2}{4a^2} \sin^2 \frac{\pi y}{b} \right] \qquad (7.31)$$

It will be noted that these equations are identical to those derived by Horne and Narayanan (1976) for plates simply supported at the edges with a sinusoidal imperfection.

Case 2

By proceeding as in the case of plates simply supported along all the four edges, the general equations for other loading edge conditions can be determined easily.

Considering an approximately square plate where loading edges are fixed at all the four edges, we may consider only a single half-wave buckled mode, i.e. $q = p = 1$.

Suitable functions for $g(x)$ and $f(y)$ would be:

$$g(x) = \left(1 - \cos \frac{2\pi x}{a}\right)$$

$$f(y) = \left(1 - \cos \frac{2\pi y}{b}\right)$$

Hence the deflection for a buckled mode of a half-wave is

$$w = A \left(1 - \cos \frac{2\pi x}{a}\right)\left(1 - \cos \frac{2\pi y}{b}\right) \tag{7.32}$$

The general equations for strain and stress are given by:

$$\alpha \, d\varepsilon_x + \frac{1}{\alpha} \, d\varepsilon_y = \frac{4}{3}\left(\frac{t}{b}\right)^2 \frac{\sigma_e}{E} \frac{dm}{m^2} \left(3\alpha^3 + \frac{3}{\alpha} + 2\alpha\right) + \frac{35}{6} \frac{\pi^2 A_0^2}{b^2} m \, dm \left(\alpha^3 + \frac{1}{\alpha}\right)$$

$$- \frac{v}{E} \left\{\frac{\alpha}{a} \int_0^a d\sigma_{ya} dx + \frac{1}{\alpha b} \int_0^b d\sigma_{xa} dy\right\} \tag{7.33}$$

$$d\sigma_{xa} = E \left\{d\varepsilon_x - \frac{3\pi^2 A_0^2}{a^2} m \, dm + \frac{v}{aE} \int_0^a d\sigma_{ya} dx\right\} \tag{7.34}$$

$$d\sigma_{ya} = E \left\{d\varepsilon_y - \frac{3\pi^2 A_0^2}{b^2} m \, dm + \frac{v}{bE} \int_0^b d\sigma_{xa} dy\right\} \tag{7.35}$$

By solving eqns (7.33), (7.34) and (7.35) numerically for the values of σ_{xa} and σ_{ya}, the stresses σ_x and σ_y at any section can be obtained from

$$\sigma_x = \sigma_{xa} + (m^2 - 1) \frac{\pi^2 E A_0^2}{a^2} \left\{\frac{3}{2} - \left(1 - \cos \frac{2\pi y}{b}\right)^2\right\} \tag{7.36}$$

$$\sigma_y = \sigma_{ya} + (m^2 - 1) \frac{\pi^2 E A_0^2}{b^2} \left\{\frac{3}{2} - \left(1 - \cos \frac{2\pi x}{a}\right)^2\right\} \tag{7.37}$$

Case 3

We will now consider an approximately square plate simply supported at $x = 0$, $x = a$, clamped at $y = 0$, $y = b$.

The suitable functions for $f(y)$ and $g(x)$ for single half-wave length would be:

$$f(y) = \left(1 - \cos \frac{2\pi y}{b}\right)$$

$$g(x) = \sin \frac{\pi x}{a}$$

Hence the deflection function is therefore:

$$w = A \sin \frac{\pi x}{a}\left(1 - \cos \frac{2\pi y}{b}\right) \tag{7.38}$$

The related equations are obtained as

$$\frac{3}{4}\,\alpha\,d\varepsilon_x + \frac{1}{\alpha}\,d\varepsilon_y = \frac{1}{4}\left(\frac{t}{b}\right)^2 \frac{\sigma_e}{E}\frac{dm}{m^2}\left(3\alpha^3 + \frac{16}{\alpha} + 8\alpha\right) + \frac{\pi^2 A_0^2}{b^2}\,m\,dm\left(\frac{35}{32}\alpha^3 + \frac{3}{2\alpha}\right)$$

$$- \frac{v}{E}\left\{\frac{3}{4}\frac{\alpha}{a}\int_0^a d\sigma_{ya}\,dx + \frac{1}{\alpha b}\int_0^b d\sigma_{xa}\,dy\right\} \tag{7.39}$$

$$d\sigma_{xa} = E\left\{d\varepsilon_x - \frac{3\pi^2 A_0^2 m\,dm}{4a^2} + \frac{v}{aE}\int_0^a d\sigma_{ya}\,dx\right\} \tag{7.40}$$

$$d\sigma_{ya} = E\left\{d\varepsilon_y - \frac{\pi^2 A_0^2 m\,dm}{b^2} + \frac{v}{bE}\int_0^b d\sigma_{xa}\,dy\right\} \tag{7.41}$$

Similarly by solving eqns (7.39), (7.40) and (7.41) numerically for the values of σ_{xa} and σ_{ya}, the stresses σ_x and σ_y at any section can be obtained from

$$\sigma_x = \sigma_{xa} + (m^2 - 1)\frac{E}{4}\frac{\pi^2 A_0^2}{a^2}\left\{\frac{3}{2} - \left(1 - \cos\frac{2\pi y}{b}\right)^2\right\} \tag{7.42}$$

$$\sigma_y = \sigma_{ya} + (m^2 - 1)\frac{E}{2}\frac{\pi^2 A_0^2}{b^2}\cos\frac{2\pi x}{a} \tag{7.43}$$

Case 4

We will now consider an approximately square plate simply supported at $x = 0$, clamped at $x = a$; simply supported at $y = 0$, clamped at $y = b$. Take

$$f(y) = \frac{1}{b^4}\left(yb^3 - 3by^3 + 2y^4\right)$$

and

$$g(x) = \frac{1}{a^4} (xa^3 - 3ax^3 + 2x^4)$$

Hence

$$w = \frac{A}{a^4 b^4} (xa^3 - 3ax^3 + 2x^4)(yb^3 - 3by^3 + 2y^4) \qquad (7.44)$$

The corresponding equations obtained are as follows:

$$\alpha \, d\varepsilon_x + \frac{1}{\alpha} \, d\varepsilon_y = \frac{1}{\pi^2} \frac{\sigma_e}{E} \left(\frac{t}{b}\right)^2 \frac{dm}{m^2} \left(21\alpha^3 + \frac{21}{\alpha} + \frac{432}{19}\alpha\right)$$

$$+ \frac{1}{58 \cdot 016} \frac{A_0^2}{b^2} m \, dm \left(\alpha^3 + \frac{1}{\alpha}\right)$$

$$- \frac{v}{E} \left\{ \frac{\alpha}{a} \int_0^a d\sigma_{ya} \, dx + \frac{1}{\alpha b} \int_0^b d\sigma_{xa} \, dy \right\} \qquad (7.45)$$

$$d\sigma_{xa} = E \left\{ d\varepsilon_x - \frac{12}{35} \times \frac{19}{630} \frac{A_0^2}{a^2} m \, dm + \frac{v}{aE} \int_0^a d\sigma_{ya} \, dx \right\} \qquad (7.46)$$

$$d\sigma_{ya} = E \left\{ d\varepsilon_y - \frac{12}{35} \times \frac{19}{630} \frac{A_0^2}{b^2} m \, dm + \frac{v}{bE} \int_0^b d\sigma_{xa} \, dy \right\} \qquad (7.47)$$

By integrating eqns (7.45), (7.46) and (7.47) numerically we obtain σ_{xa} and σ_{ya}; the stresses σ_x and σ_y at any section can be obtained from

$$\sigma_x = \sigma_{xa} + (m^2 - 1) \frac{6}{35} \frac{EA_0^2}{a^2} \left[\frac{19}{630} - \left\{ \frac{1}{b^4} (yb^3 - 3by^3 + 2y^4) \right\}^2 \right] \qquad (7.48)$$

$$\sigma_y = \sigma_{ya} + (m^2 - 1) \frac{6}{35} \frac{EA_0^2}{b^2} \left[\frac{19}{630} - \left\{ \frac{1}{a^4} (xa^3 - 3ax^3 + 2x^4) \right\}^2 \right] \qquad (7.49)$$

Case 5

We will now consider an approximately square plate clamped at $x = 0$, $x = a$, simply supported at $y = 0$, clamped at $y = b$. The suitable functions for $f(y)$ and $g(x)$ would be:

$$f(y) = \frac{1}{b^4} (yb^3 - 3by^3 + 2y^4)$$

$$g(x) = \left(1 - \cos \frac{2\pi x}{a}\right)$$

Thus the deflection function is

$$w = \frac{A}{b^4} \left(1 - \cos\frac{2\pi x}{a}\right)(yb^3 - 3by^3 + 2y^4) \qquad (7.50)$$

The equations are given by

$$\alpha\,d\varepsilon_x + \frac{0\cdot8639}{\alpha}\,d\varepsilon_y = \left(\frac{t}{b}\right)^2 \frac{\sigma_e}{E}\frac{dm}{m^2}\left(4\alpha^3 + \frac{1\cdot8382}{\alpha} + 2\cdot3037\alpha\right)$$

$$+ \frac{A_0^2}{b^2}\,m\,dm\left(0\cdot9917\alpha^3 + \frac{0\cdot8639}{\alpha}\right)$$

$$- \frac{v}{E}\left\{\frac{\alpha}{a}\int_0^a d\sigma_{ya}\,dx + \frac{0\cdot8639}{b\alpha}\int_0^b d\sigma_{xa}\,dx\right\} \quad (7.51)$$

$$d\sigma_{xa} = E\left\{d\varepsilon_x - \frac{19}{315}\frac{\pi^2 A_0^2}{a^2}\,m\,dm + \frac{v}{aE}\int_0^a d\sigma_{ya}\,dx\right\} \qquad (7.52)$$

$$d\sigma_{ya} = E\left\{d\varepsilon_y - \frac{18}{35}\frac{A_0^2}{b^2}\,m\,dm + \frac{v}{bE}\int_0^b d\sigma_{xa}\,dy\right\} \quad (7.53)$$

As before, integrating eqns (7.51), (7.52) and (7.53) numerically for σ_{xa} and σ_{ya}, the stresses σ_x and σ_y at any section can be obtained:

$$\sigma_x = \sigma_{xa} + (m^2 - 1)E\,\frac{\pi^2 A_0^2}{a^2}\left[\frac{19}{630} - \left\{\frac{1}{b^4}(yb^3 - 3by^3 + 2y^4)\right\}^2\right] \qquad (7.54)$$

$$\sigma_y = \sigma_{ya} + (m^2 - 1)\frac{6}{35}\frac{EA_0^2}{b^2}\left[\frac{3}{2} - \left(1 - \cos\frac{2\pi x}{a}\right)^2\right] \qquad (7.55)$$

Case 6

We will now consider an approximately square plate simply supported at $x = 0$, $x = a$ and $y = 0$, and clamped at $y = b$.

The suitable functions for $f(y)$ and $g(x)$ would be:

$$f(y) = \frac{1}{b^4}(yb^3 - 3by^3 + 2y^4)$$

$$g(x) = \sin\frac{\pi x}{a}$$

Thus the deflection function is

$$w = \frac{A}{b^4} \sin \frac{\pi x}{a} \, (yb^3 - 3by^3 + 2y^4)$$

The equations are given by

$$\frac{19}{36} \pi^2 \alpha \, d\varepsilon_x + \frac{6}{\alpha} \, d\varepsilon_y = 7 \frac{\sigma_e}{E} \left(\frac{t}{b}\right)^2 \frac{dm}{m^2} \left(\pi^2 \frac{19}{252} \alpha^3 + \frac{18}{\pi^2} \frac{1}{\alpha} + \frac{12}{7} \alpha\right)$$

$$+ \frac{A_0^2}{b^2} \, m \, dm \left(1 \cdot 292 \alpha^3 + \frac{54}{35} \frac{1}{\alpha}\right)$$

$$- \frac{\nu}{E} \left\{ \frac{19}{36} \pi^2 \frac{\alpha}{a} \int_0^a d\sigma_{ya} \, dx + \frac{6}{b\alpha} \int_0^b d\sigma_{xa} \, dy \right\}$$

$$\tag{7.56}$$

$$d\sigma_{xa} = E \left\{ d\varepsilon_x - \frac{19}{1260} \frac{\pi^2 A_0^2}{a^2} \, m \, dm + \frac{\nu}{aE} \int_0^a d\sigma_{ya} \, dx \right\}$$

$$\tag{7.57}$$

$$d\sigma_{ya} = E \left\{ d\varepsilon_y - \frac{6}{35} \frac{A_0^2}{b^2} \, m \, dm + \frac{\nu}{bE} \int_0^b d\sigma_{xa} \, dy \right\} \tag{7.58}$$

As before, integrating eqns (7.56), (7.57) and (7.58) numerically for σ_{xa} and σ_{ya}, the stresses σ_x and σ_y at any section can be obtained:

$$\sigma_x = \sigma_{xa} + (m^2 - 1)E \frac{\pi^2 A_0^2}{4a^2} \left[\frac{19}{630} - \left\{ \frac{1}{b^4} \, (yb^3 - 3by^3 + 2y^4) \right\}^2 \right]$$

$$\tag{7.59}$$

$$\sigma_y = \sigma_{ya} + (m^2 - 1) \frac{3}{35} \frac{EA_0^2}{b^2} \cos \frac{2\pi x}{a} \tag{7.60}$$

7.4   STRESS CRITERION

As the state of stress in the plate is biaxial, rather than uniaxial, it is necessary to define a limiting stress criterion which incorporates the two stress components acting on an element of the plate material. This criterion must be adequate to determine that the plate element has yielded.

It is usual to employ the following semi-empirical stress criterion proposed by Von Mises for ductile materials.

$$\sigma_{eq} = (\sigma_x^2 + \sigma_y^2 - \sigma_x \sigma_y)^{1/2}$$

An element is considered to yield when the equivalent stress reaches the yield stress of the material in uniaxial tension (i.e. $\sigma_{eq} = \sigma_{ys}$).

However, the analysis described in the foregoing pages is based on the Energy method with an assumed deflection surface; such an analysis is known to give an upper-bound solution. To allow approximately for this, a similar but conservative yield criterion proposed by Haigh and Beltrami (see Timoshenko, 1960) has been adopted in this analysis:

$$\sigma_{eq} = (\sigma_x^2 + \sigma_y^2 - 2\nu\sigma_x\sigma_y)^{1/2}$$

As has been stated before, very little theoretical work has been done on biaxially loaded plates and fewer experimental results have been published. Chow (1983) has compared the theoretical predictions using the above analysis with experiments carried out by him and has found that the method gives satisfactory estimates of mean stress at collapse. He has also compared the results obtained by the more exact dynamic relaxation finite difference method (Coombs, 1975) with predictions using the above theory. The range of parameters studied included various b/t ratios, applied strain ratios and selected degrees of initial imperfection for the square plates. In all cases, he found the predictions to be sufficiently accurate.

7.5   FACTORS AFFECTING THE STRENGTHS OF BIAXIALLY LOADED PLATES

As has been mentioned before, the complicating factors of the biaxial compression problem are the marked effects of plate slenderness, aspect ratio, initial geometric imperfection (shape and amplitude) and boundary conditions on buckling stiffness and strength.

Aspect Ratio
Table 7.2 compares the results obtained on square plates and plates with aspect ratio $(a/b) = 3$ having the same initial bows and slenderness ratio of $b/t = 60\cdot0$. Both types of plates are simply supported along all four edges. For equal average strains applied in the orthogonal directions, the longitudinal strengths of such a rectangular plate with a single half-wave initial bow are greater than those for a square plate with the same magnitude of initial deformation; however, the transverse peak strengths are in the reverse order. This is also true when the applied strain ratio is doubled.

TABLE 7.2

COMPARISON BETWEEN 3:1 AND 1:1 PLATES SIMPLY SUPPORTED ALONG ALL EDGES

No. of half-waves (p)		1 3:1 Plate		2 Square plate		3 3:1 Plate	
δ_0 (mm)	$\dfrac{\varepsilon_x}{\varepsilon_y}$	$\dfrac{\sigma_x}{\sigma_{ys}}$	$\dfrac{\sigma_y}{\sigma_{ys}}$	$\dfrac{\sigma_x}{\sigma_{ys}}$	$\dfrac{\sigma_y}{\sigma_{ys}}$	$\dfrac{\sigma_x}{\sigma_{ys}}$	$\dfrac{\sigma_y}{\sigma_{ys}}$
3	1·0	0·65	0·20	0·43	0·43	0·43	0·43
0·75	1·0	0·64	0·23	0·49	0·49	0·49	0·49
3	2·0	0·80	0·17	0·59	0·23	0.59	0·23
0·75	2·0	0·77	0·23	0·65	0·32	0·65	0·32

$$E = 205\,000 \text{ N/mm}^2 \qquad q = 1 \qquad a = 720 \text{ mm}$$
$$\sigma_{ys} = 245 \text{ N/mm}^2 \qquad\qquad b = 240 \text{ mm}$$
$$b/t = 60 \cdot 0$$

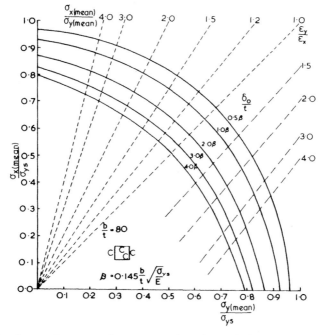

FIG. 7.2.   Interaction curves for a square plate clamped along all edges, $b/t = 80$.

If the initial imperfection for the above rectangular plate is assumed to have three half-waves longitudinally, the strength of the plate in both the long and short direction is identical to the strength of the square plate; this is obviously due to the fact that each square portion of the 3:1 plate has the same dimension and loading conditions as the square plate, provided the edges are constrained to remain straight and have a constant displacement applied along each edge.

Initial Imperfections

The interaction curves for a square plate of $b/t = 80$ clamped along all the four edges and having various degrees of initial imperfection are shown in Fig. 7.2. As in the case of uniaxial compression, the peak strength reduces with an increase in initial imperfection.

Figure 7.3 shows the plot of longitudinal stress and corresponding transverse stress for a square plate in which the initial imperfection is given by $\delta_0/t = 0.145(b/t)\sqrt{(\sigma_{ys}/E)}$; the plate is simply supported along all the

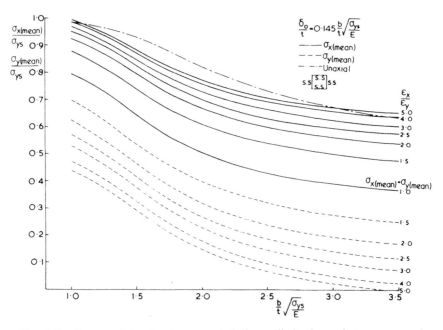

FIG. 7.3.   Square plates simply supported along all the four edges $\sigma_{x(mean)}$ and $\sigma_{y(mean)}$ for different $\varepsilon_x/\varepsilon_y$ ratios.

four edges. (The initial imperfection given by this value of δ_0 is considered to be adequate for all practical plates (see Horne, 1980).) The results of uniaxial strength are included for comparison. For a strain ratio $\varepsilon_x/\varepsilon_y >$ 4·0, and b/t ratio less than 30, the strengths are marginally higher than the uniaxial strengths. Hence such biaxially loaded plates could be treated as uniaxially loaded plates when they are stocky (b/t less than 30). When b/t is greater than 30 (i.e. more slender plates) such simplification is no longer valid.

7.6 DESIGN INTERACTION CURVES

The interaction curves for square plates having initial imperfection $\delta_0/t = 0·145(b/t)\sqrt{(\sigma_{ys}/E)}$ for six combinations of boundary conditions are presented in Figs. 7.4–7.9. The longitudinal mean stress at collapse (σ_x) is plotted against the transverse mean stress at collapse (σ_y) for various σ_x/σ_y, $\varepsilon_y/\varepsilon_x$ and b/t ratios, so that the corresponding longitudinal and

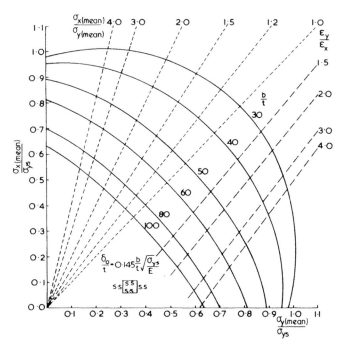

FIG. 7.4.   Interaction curves for square plates simply supported along all edges.

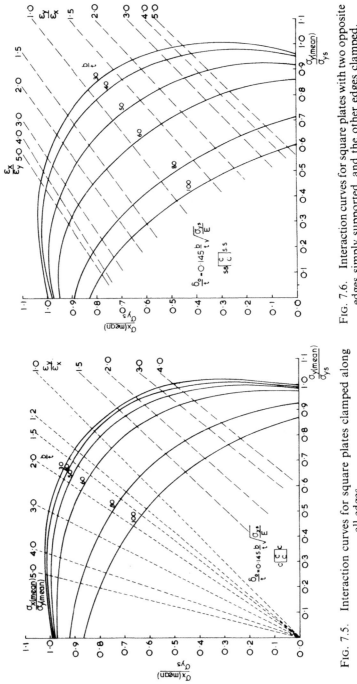

FIG. 7.5. Interaction curves for square plates clamped along all edges.

FIG. 7.6. Interaction curves for square plates with two opposite edges simply supported, and the other edges clamped.

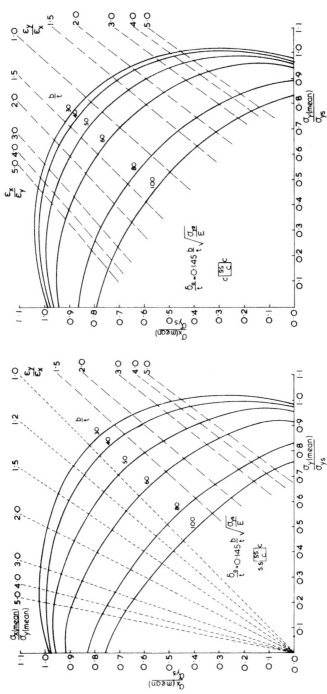

FIG. 7.7. Interaction curves for square plates with two adjacent sides simply supported, and the other edges clamped.

FIG. 7.8. Interaction curves for square plates with three adjacent sides clamped, and the fourth simply supported.

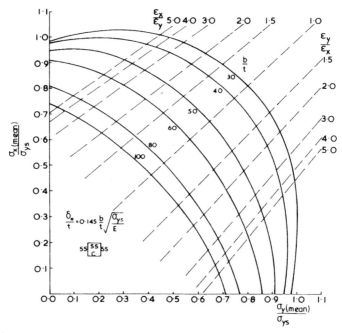

FIG. 7.9.   Interaction curves for square plates with three adjacent sides simply supported, and the fourth edge clamped.

transverse stresses can be obtained directly from the curves. (The abscissae and the ordinates are plotted in a non-dimensional scale, by dividing the mean stress σ_x and σ_y by the yield stress σ_{ys}.) The b/t ratios used ranged from 30–100, and are within the practical limits.

7.7 CONCLUSION

An approximate method to predict the post-buckling and collapse behaviour of plates subjected to biaxial loading is presented. The analysis is based on the energy method and accounts for the effect of initial imperfections and the ratios of applied strain in the two directions; the effect of aspect ratio has also been discussed. The suggested analysis is adequate to predict the ultimate compressive strengths of biaxially loaded plates. Interaction curves have been presented for the use of designers and incorporate various boundary conditions and a wide range of b/t ratios.

ACKNOWLEDGEMENT

The computational assistance given by Mr F. Y. Chow is hereby gratefully acknowledged.

REFERENCES

BECKER, H., GOLDMAN, R. and POZERYCKI, J. (1970) Compressive strength of ship hull girders, part 1, unstiffened plates. Ship Structures Committee Technical Report SSC 217 on Small Hull Girder Model, US Coastguard Headquarters, Washington, DC.

CHOW, F. Y. (1983) Plates under axial compression and shear. PhD Thesis, University of Wales.

COOMBS, M. L. (1975) Aspects of the elasto-plastic behaviour of biaxially loaded plates. MSc Thesis, University of London.

CRISFIELD, M. A. (1973) Large deflection elasto-plastic buckling analysis of plates using finite elements. Transport and Road Research Laboratory, Crowthorne, Report No. LR 593.

DOWLING, P. J. (1974) Some approaches to the non-linear analysis of plated structures. *Symposium on Non-linear Techniques and Behaviour in Structural Analysis*, Transport and Road Research Laboratory.

HORNE, M. R. (1980) Basic concepts in the design of webs. *Proceedings of the Cardiff Conference on the New Code for the Design of Steel Bridges*, Paper No. 10.

HORNE, M. R. and NARAYANAN, R. (1976) Strength of axially loaded stiffened panels. *Memoires of the International Association for Bridge and Structural Engineering*, Zurich, pp. 125–57.

NARAYANAN, R. and SHANMUGAM, N. E. (1980) Effective widths of axially loaded plates. *Journal of Civil Engineering Design*, 1(3), 253–72.

SMITH, C. S. (1975) Compressive strength of welded steel ship grillages. *RINA Spring Meetings*, Paper No. 9.

TIMOSHENKO, S. (1936) *Theory of Elastic Stability*, McGraw-Hill, New York.

TIMOSHENKO, S. (1960) *Strength of Materials, Part 2*, 3rd Edition, Van Nostrand, New York.

WILLIAMS, D. G. (1971) Some examples of the elastic behaviour of initially deformed bridge panels. *Civil Engineering and Public Works Review*.

WILLIAMS, D. G. (1975) The analysis of initially deformed plates subjected to uniaxial and biaxial compression. British Ship Research Association, Report NS426.

YAM, L. P. C. (1974) Finite difference method for non-linear plate problems and parametric study. *Symposium on Non-linear Techniques and Behaviour in Structural Analysis*, Transport and Road Research Laboratory, Crowthorne.

ZIENKIEWICZ, O. C., VALLIAPPAN, S. and KING, I. P. (1969) Elastoplastic solutions of engineering problems, initial stress, finite element approach. *International Journal of Numerical Methods in Engineering*, pp. 75–100.

Chapter 8

THE INTERACTION OF DIRECT AND SHEAR STRESSES ON PLATE PANELS

J. E. HARDING

Imperial College of Science and Technology, London, UK

SUMMARY

Plates under various types of loading, such as uniaxial compression, biaxial compression, shear, etc., are studied. Various parameters which affect their stability and strength are considered in detail. Computer studies on the interaction between in-plane compression, tension and bending stresses are presented. Interaction formulae used in design codes are discussed.

8.1 INTRODUCTION

Before considering the behaviour of stiffened plates under complex loading conditions it is essential to understand the basic phenomena controlling plate behaviour and their importance and effect.

Historically, much early work concentrated on the study of the critical buckling of plates and it is only in recent years that the numerical analysis of plates including both geometric and material non-linearities has become possible. It is towards this latter behaviour that this chapter is devoted.

Over the last ten years, computer studies, both finite difference and finite element as well as other approaches, have made great strides in the understanding of complex plate behaviour and in particular a knowledge of the ultimate strength and post-collapse stiffness of plates with fabrication imperfections and material non-linearity. Work in Great Britain at Cambridge University, the Road Research Laboratory at Crowthorne, and at Imperial College of Science and Technology in London has led the world in an understanding of the behaviour of unstiffened plates. The change in the British codes of practice to a limit-state design philosophy has prompted or at least spurred on much of this effort.

This chapter deals with some of the basic phenomena relevant to plate behaviour, illustrates these in terms of simple loading types, and then progresses into the more complex interactive situation. It deals particularly with results produced at Imperial College using finite difference large deflection elasto-plastic computer packages.

8.2 NON-LINEARITY IN PLATE BEHAVIOUR

The theory of plates began with an enormous volume of work devoted to the critical buckling of plates under various loading types (Bleich, 1982).

The critical buckling load of a plate is essentially that load at which the plate becomes unstable under the action of its in-plane loading. When subjected to some minor perturbation the plate will tend to snap into another path which will often include deflection components not associated with the original loading. A plate loaded in uniaxial compression, for example, will reach a limit load whereby, if the plate were initially perfect, the behaviour will change from an in-plane stress–strain response to a more complex behaviour involving out-of-plane deflections. The behaviour of the plate after the critical load may be stable or unstable. In the case of a plate under uniaxial compression the post-buckling behaviour will be stable because of the membrane tensions associated with out-of-plane deflections.

This idealised view of plate behaviour, ignores the initial factors that occur in any practical situation. The first of these is the presence of imperfections in the plate, in general terms always present because of the impossibility of a naturally perfect configuration but in particular terms present to a significant degree in civil engineering structures because of the relative crudity of the fabrication processes. These processes, in particular welding associated with stiffeners or plate boundaries, build in residual stresses and geometric imperfections which when combined with a non-linear material characteristic, the second of the important practical considerations, have an effect on plate yield and therefore an effect on pre- and post-collapse stiffness and ultimate load.

The presence of a non-linear material characteristic, and in particular with structural steels the presence of a yield point, mean that for plates whose critical stress is higher than their yield stress (allowing for out-of-plane bending effects, etc., on yield), yielding occurs which results in loss of stiffness and collapse strength well below the critical stress.

Because imperfections cause an increase in the growth of out-of-plane deflections as loading progresses they will have an effect on the degree of plate bending stresses present and thereby cause earlier yielding with resultant general effects. This behaviour can be summarised by reference to Fig. 8.1 which shows an approximate ultimate load versus slenderness curve for a plate panel subjected to uniaxial compression.

The curve is divided into three regions which separate out to some degree the main parameters controlling failure.

In region A the behaviour of the plate is dominated by the material yield stress. The yield point is well below the critical buckling stress and the latter has virtually no influence on behaviour. As a result the plate behaviour is controlled by in-plane actions and geometrical imperfections have little effect. Residual stresses do have an influence on pre-collapse stiffness by causing early in-plane yield and this effect will be seen in more detail in the next section but because of the lack of effect of out-of-plane instability, the plate is able to redistribute these residual stresses and they have negligible effect on the ultimate collapse load. The collapse load of the plate will always, for practical cases, be the yield stress of the material. This ignores the effect of strain hardening which is commonly neglected for steel with a yield plateau.

Region B is the area where both material yield and critical buckling interact and both have direct influence on the behaviour. As a result, both

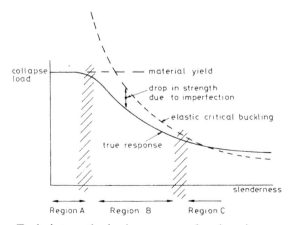

FIG. 8.1.   Typical strength–slenderness curve for plates in compression.

geometric imperfections, which cause growth of deformation, loss of stability and resulting early yielding, and residual stresses which cause early yielding and resulting loss of stiffness, have an important influence on pre- and post-collapse behaviour and peak load. At the point where the yield stress and critical buckling stress of a plate coincide ($b/t = 55$ for a mild steel plate with $\sigma_0 = 245 \, \text{N/mm}^2$) the sensitivity to imperfections is at a maximum and the biggest difference occurs between the idealised lower bound of yield stress and critical buckling stress and true plate collapse. It is also the region where a large scatter of experimental results is obtained because of imperfection sensitivity.

Region C is the area dominated by elastic critical buckling. The critical buckling stress of the plate is significantly less than the yield stress and very large out-of-plane deflections are needed before yield occurs. Yield therefore, even with residual stresses of moderate orders present, has little effect on pre-collapse stiffness and collapse load. It can be seen from Fig. 8.1 that the plate collapse strength is greater than the critical stress and this is because of post-buckling stability resulting from the transverse in-plane tensions set up by the growth of out-of-plane deflections.

Some or all of the effects detailed above are present for other types of loading such as shear or for loading types in combination. For other loading types the relative plate stockiness is different (the critical buckling stress is different for a given slenderness) and the degree of imperfection sensitivity, etc., is modified. These effects are discussed in more detail in the following sections.

8.3 PLATES UNDER UNIAXIAL COMPRESSION

8.3.1 Introduction
Frieze *et al.* (1977) is the most comprehensive study conducted to date on the elasto-plastic large deflection behaviour of plates subjected to uniaxial compression. The results of the study illustrate the effect of initial deformation, residual stress, plasticity, degree of boundary restraint, aspect ratio and plate slenderness.

8.3.2 Effect of Aspect Ratio
The first part of the study concentrated on the effect of aspect ratio and concluded that the behaviour of a square plate generally gave a conservative estimate of strength and stiffness for long plate panels of the type found in bridge structures.

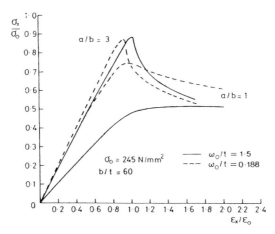

FIG. 8.2.   Effect of aspect ratio on plates in compression.

This is illustrated in Fig. 8.2 which shows the average stress–strain characteristic for a mild steel plate with a breadth to thickness ratio (b/t) of 60 and two levels of initial imperfection. In the case of both plates the imperfection has been taken as a single wave in both directions. This results in a peakier response for the 3:1 plate because of later surface yielding with a fast unloading characteristic induced by snap-through buckling into a three-wave mode. If the initial imperfection for the 3:1 plate had been in a three half-wave longitudinal mode the response would have approximated to that of the square plate. In reality a component of both these and other modes would be present inducing a strength somewhere between the two extremes.

The effect of the imperfection can be clearly seen both in its reduction of the pre-collapse plate stiffness and in the case of the square plate, its reduction in peak strength. The latter is not noted for the long plate because of the non-sympathetic collapse mode. Because of the conservatism of the approach all subsequent studies conducted were on plates of square aspect ratio.

8.3.3 Loaded Edge Restraint
The curves of Fig. 8.2 correspond to the weakest degree of transverse boundary restraint. The 'unrestrained' condition used in this study corresponds to the unloaded plate edges carrying zero transverse in-plane stress and zero boundary in-plane shear. Two other conditions were studied. These were 'restrained' edges where the unloaded edges remained

in position and were capable of carrying transverse stresses and a 'net' transverse force, and the 'constrained' condition where the unloaded edges remained straight but where allowed to move, could carry transverse stresses but no net transverse force.

The relative effect of these edge conditions is shown in Fig. 8.3. For the stocky plate ($b/t = 20$) the unrestrained and constrained characteristics are very similar. This is because the restrained boundary stresses are fairly uniform in this case and allowing the boundary to move to relax the net transverse force results in very low stresses approximating to the unrestrained condition. The restrained strengths are higher and a Poisson effect allows stresses significantly in excess of uniaxial yield. For the slender plate the restrained transverse boundary stresses are highly non-uniform because of the presence of large out-of-plane deflections, being tensile at the centre of the panel edge and compressive near the corners. Relaxation to a zero net force causes little change in boundary stresses and it can be seen that the constrained-edge condition response approaches that of the restrained.

As this study was aimed at the stiffened compression flange of a box-girder bridge subsequent analyses concentrated on the constrained condition. It was considered that the degree of edge restraint was insufficient, because of the presence of the virtually free flange edge, to allow a restrained condition while the presence of repeated adjacent panels kept the edge relatively straight. The subsequent modification of this approach to design is discussed later.

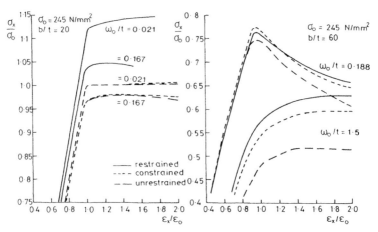

FIG. 8.3.   Effect of edge restraint on plates in compression.

8.3.4 Effect of Geometric Imperfections

Figure 8.4 shows the average stress–strain characteristics of a panel with a medium slenderness ($b/t = 60$ for $\sigma_0 = 245\,\text{N/mm}^2$) subjected to various levels of initial geometrical imperfections but with zero applied residual stress. The parameter shown on the figure is of interest as it has been demonstrated to be the non-dimensionalising parameter with respect to yield stress. The ω_0' parameter ($=\omega_0 t$) is the non-dimensional initial imperfection with respect to plate thickness.

It should be noted that the design-tolerance level accepted in the UK ($\omega_0 = 0\cdot145\beta t$) corresponds to a ω_0' of 0·3 in this case. This level is generally considered appropriate for normal fabrication.

It can be seen from the curves of Fig. 8.4 that while strength reductions are still occurring for unrealistic levels of imperfection as large as one and a half times the plate thickness, the majority of the weakening effect within the practical range has occurred inside this design tolerance. It should also be noted that the combination of this tolerance level and a 10 % residual stress (see next section) produces a sufficient level of strength reduction to cope with most fabrication defects from a 'good standard' fabrication yard. This is discussed in Dowling *et al.* (1977) which concludes that by adjusting a fairly generous plate tolerance ($\omega_0 = 0\cdot145\beta t$) losses of strength are a maximum of about 10 % greater than achieved by tighter tolerances suggested elsewhere and that the experience of UK fabrication suggests that such a tolerance can easily be met using normal fabrication techniques. The main benefit of adopting such a tolerance is that reliance can be based on visual inspection in the shop, thus reducing the need for close inspection.

Reference to Fig. 8.4 shows the reduction in 'elastic' stiffness and the

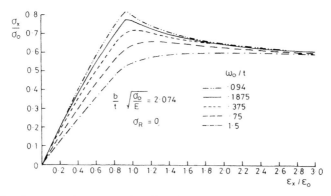

FIG. 8.4.   Stress–strain response for plates in compression—effect of imperfection.

change in post-collapse stiffness with increasing imperfection. The former has implications on complex multi-element stiffened structures as it may have a detrimental effect on stiffened panel collapse. The latter shows clearly the 'softening' effect of imperfections removing the sudden failure mechanisms with rapid growth of out-of-plane deflections. The curves also illustrate that the ultimate strength (strength at high strain) approaches a consistent value once the out-of-plane deflections have grown sufficiently in all cases.

8.3.5 Effect of Residual Stress

Figure 8.5 shows the effect of adding residual stresses to a plate of moderate slenderness ($b/t = 40$ for $\sigma_0 = 245\,\mathrm{N/mm^2}$) and a level of imperfection approximately equal to the design tolerance.

The type of residual stress distribution considered is shown in Fig. 8.6. It basically consists of high tensile strains induced in the area of the weld by plastic flow and shrinkage of the weld metal on cooling. This tensile zone induces lower compressive stresses in the rest of the panel and it is these stresses, which combined with the applied compression, produce early yield. These stresses are modified to some degree by the plate slenderness (the inability to carry a uniform compression) and to some degree by the computer idealisation. Figure 8.6 also shows the modified stress distribution for a plate of $b/t = 60$ with a high level of residual stress.

The effect of the residual stress on the response curves shown in Fig. 8.5

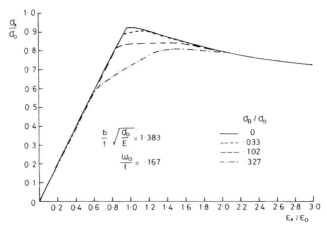

FIG. 8.5. Stress–strain response for plates in compression—effect of residual stress.

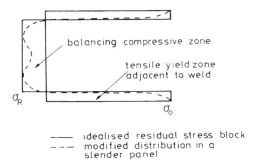

balancing compressive zone

tensile yield zone
adjacent to weld

σ_R

σ_o

—— idealised residual stress block
– – – modified distribution in a
        slender panel

FIG. 8.6.   Analytical residual stress distributions.

can be clearly seen. The biggest effect is on 'elastic' stiffness once surface yielding has recurred at the plate centreline. A residual stress level of 0·327 of yield in compression causes first yield at 0·6 of the yield stress compared with about 0·9 for a plate with zero residual stress. For plates of moderate slenderness the first yield reduction corresponds closely to the level of compressive residual stress. It can be noted, however, that in spite of this drastic loss of stiffness the effect on ultimate strength is only moderate. As the stockiness increases and the unloading curve becomes flatter the effect of residual stress on peak load becomes still less marked although the peak can occur at quite high strain levels. This again may have implications on the stability of the structure as a whole.

An important feature of imperfections should be noted here. Essentially, the effect of geometrical imperfections and residual stresses are not additive. If a geometrically perfect plate were subjected to the levels of residual stress considered here, the effect on peak stress, in particular, would be far more marked. The presence of realistic fabrication imperfections, however, removes a large part of this strength reduction and subsequent losses are relatively small. Early studies on the effect of residual stresses conducted on near-perfect plates produced very misleading conclusions about their detrimental effects.

The variation with plate slenderness of the effects discussed in this section can be seen in Fig. 8.7. These ultimate strength curves have been drawn with an imperfection level ω_0/t of $0·087\beta^2$ (an earlier version of the design tolerance). This tolerance becomes approximately equal to the tolerance of $0·145\beta$ at a plate slenderness of 60. The curves show that strength is most affected by residual stress at intermediate plate slenderness values as stated previously. The lightly welded and heavily welded curves approximate to compressive residual stress levels of 10 % and 30 % of yield. The maximum

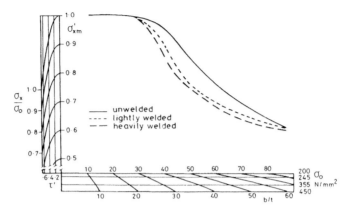

FIG. 8.7.   Strength–slenderness curves for plates in compression.

strength reduction due to the lower level is of the order of 10 %. The curves also show approximate strength reductions due to the presence of coexistent shear obtained from the study discussed in Section 8.6. They should only be used in situations where compression dominates the behaviour and shear is of secondary importance such as in the compression flange of a girder. The values given are average panel shear stresses divided by shear yield stress.

While the discussion of this section has concentrated on the results of one study, other studies (Moxham, 1971; Little and Dwight, 1972; Crisfield, 1975) have produced results which illustrate the same trends and come to the same general conclusions.

8.4   PLATES UNDER BIAXIAL COMPRESSION

8.4.1   Introduction
A similar parametric study to that applied in Section 8.3 has been used to study the behaviour of plates subjected to various combinations of longitudinal and transverse compressive displacements (Dowling *et al.*, 1979). Valsgard (1978) provides useful data applicable to the same problem.

Attention was concentrated in the work of Dowling *et al.* (1977) on plates with a 3:1 aspect ratio. This was largely because the application of the work was to ships' decking which is largely composed of panels of high aspect ratio. A consideration of square panels would overestimate the transverse compressive strength.

Imperfections considered were of a two-mode type incorporating both one and three half-wave modes in the longitudinal direction with the latter having a peak amplitude of one-half the single mode. Three levels were chosen with the average level having a single half-wave amplitude $\omega_{01}/t = 0.1\beta^2$ and a three half-wave amplitude $\omega_{03}/t = 0.05\beta^2$. Slight and severe imperfection levels of one-quarter and three times these values were also considered.

Compressive residual stress levels of 5%, 20% and 40% of yield were considered in the longitudinal direction while levels of only 0.8%, 2.9% and 5% were taken in the transverse direction. Low transverse levels were adopted because of the wide spacing of transverse stiffeners. A wide range of plate slenderness (b/t) from 20 to 120 was considered. Loading was applied by means of biaxial uniform displacements with sufficient cases being run to build up peak stress interaction diagrams of σ_x against σ_y (the x-direction being the longitudinal direction of the plate).

8.4.2 Effect of Loading Ratio on Stress–Strain Response

Figure 8.8 shows stress–strain responses for plates with b/t values of 20, 50 and 120 with the average levels of geometric and material imperfections quoted above.

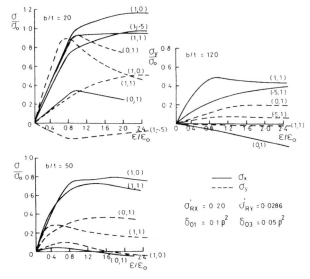

FIG. 8.8.   Stress–strain characteristics of plate under biaxial compression.

The solid curves show the σ_x response while the dotted curves show the σ_y response. In all cases the strain axis represents the highest average strain applied to the plate (either ε_x or ε_y). The numbers in brackets represent the applied strain ratio, e.g. (1, 0) corresponds to loading only with $\varepsilon_y = 0$, and (1, −0·5) represents $\varepsilon_y = -0.5\varepsilon_x$ (i.e. tensile loading). The latter points were needed in some instances to complete the interaction diagrams. The solid and dotted curves corresponding to (1, 0) in the $b/t = 20$ case, for example, therefore represent coincident response of a plate under longitudinal compressive displacement only.

It is difficult to quantify the effects in Fig. 8.8 as the behaviour is complex. In general terms the relatively sharp unloading of the σ_y stresses can be seen although the different ratios of the ε_y-axis should be remembered. For example, for the (0, 1) and (1, 1) cases, the ε-axis corresponds to ε_y, while for the case (1, −0·5), the ε-axis corresponds to $2\varepsilon_y$.

This greater slope of the σ_y unloading curves compared with those of σ_x reflects the lack of stability of long plates loaded transversely and indicates the lack of sympathy of the ω_{03} component of the imperfection mode to collapse mode.

One important point to note here is that results have been obtained in this study for simply-supported boundary conditions. This is a lower-bound approach normally thought appropriately conservative for plate analysis where the degree of clamped restraint afforded by symmetric 'hungry horse' imperfections induced by stiffener welding cannot be relied on. In most instances the effect only has a small influence, but in the case of transverse loading of long panels, the added benefit from boundary clamping can appreciably increase peak load. In Smith (1981) this effect is demonstrated and Fig. 8.9 reproduced from that paper shows the difference

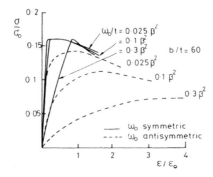

Fig. 8.9.   Effect of imperfection mode on plates under biaxial compression.

between strength produced by antisymmetrical imperfections (effectively producing a simply-supported boundary) and symmetrical imperfections (effectively producing a clamped boundary).

The use that can be made of this degree of clamping in practical design must depend on the application, type of fabrication and level of safety required. In bridge structures where the effect is likely to be unpredictable and benefits are generally low it seems less desirable to take account of the out-of-plane restraint, while for ships where a significant proportion of the loading is transverse to the panel and where service actions as well as fabrication tend to enforce the 'hungry horse' shape a degree of benefit may be possible.

The remaining results in this section, however, adopt the simply-supported edge restraint for conservatism.

8.4.3 Interaction Results

A difficulty arises in the case of interactive loading in the production of a unique peak load interaction. A similar problem exists for the interaction of shear and direct stress and this will be discussed later.

With biaxial compression two peak stress-interaction curves are possible, namely peak σ_x against coincident σ_y, and peak σ_y against coincident σ_x. The two curves are significantly different in many cases. The problem arises because the application of load via displacements (normally the simplest and most rewarding approach in numerical approaches) does not require that the two stress components peak simultaneously.

The problem is illustrated in Fig. 8.10. For stocky plates (Fig. 8.10(a), $b/t = 20$ and 30) the maximum σ_x curve forms a loop below the maximum σ_y curve and rejoins it. The sketch showing the progressive relationship between σ_x and σ_y as the displacements are applied illustrates this phenomenon. In this case, as strain is applied, σ_y peaks first, while σ_x is still rising. With increasing applied strain σ_y begins to fall while σ_x eventually reaches a peak. This behaviour is shown as one of the dotted lines in the main part of Fig. 8.10(a) and two distinct interaction curves result. It can be argued that the outer curve always corresponds to a strain condition that the plate can sustain and can therefore always be adopted but caution should be exercised in this approach.

The outer curve may correspond to a situation where one of the two stresses is shedding and the remainder of the structure may not be able to sustain this. If the true loading situation is one of applied stress it seems likely that the resulting interaction will be between the two extremes although it could well approach the upper curve as being the limit situation.

The precise nature of the type of structure and loading, however, will influence the type of behaviour that results.

It should be noted that for stocky plates Fig. 8.10 shows that the outer curve is always obtained with both curves rising and it would therefore seem quite reasonable to adopt this for design. For more slender plates, however, when one of the stresses predominates, the peak stress for the weaker

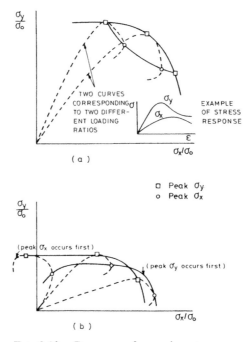

FIG. 8.10.   Response of σ_x and σ_y stresses.

component occurs significantly before that of the stronger component and the diagram in Fig. 8.10(b) illustrates this. In this instance unloading is occurring in the case of the outside curve and to adopt this curve would at the very least be ignoring some of the potential panel strength in the weaker direction. A design philosophy for the structure as a whole therefore needs to be adopted in this type of interactive situation and testing is needed under applied stress control to confirm the resulting behaviour. A degree of moderation seems justified in design.

8.4.4 Variation in Strength with Slenderness

Figure 8.11 shows the resulting interaction curves for average levels of geometric and material imperfections. The two branches of the curves are also shown.

For the stocky panels a biaxial effect induces stress levels greater than yield in certain areas while the circular yield interaction controls general behaviour. For the slender panels behaviour is much closer to a straight line reflecting elastic buckling influences.

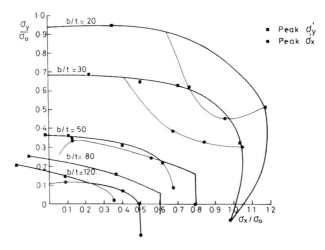

FIG. 8.11.   Interaction curves for plates under biaxial compression.

8.4.5 Effect of Imperfection

Figure 8.12 shows the effect of geometrical imperfection on the strength of plates with b/t values of 20, 50 and 120. The biaxial stress state reduces the imperfection critical b/t of 55 in the case of uniaxial compression and the $b/t = 20$ curves have a higher sensitivity to imperfection in absolute terms than the $b/t = 50$ set. The $b/t = 30$ case, not shown, has the biggest sensitivity of all the slenderness values studied with peak σ'_y strength ranging from 0·97 to 0·51.

The effect of the residual stress level is much less marked, as for the case of uniaxial compression.

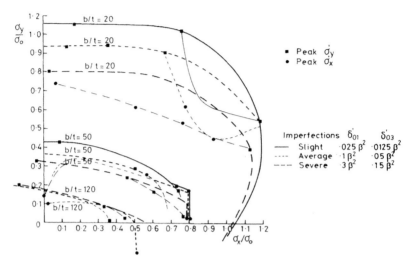

FIG. 8.12.   Interaction curves for plates under biaxial compression—effect of imperfections.

8.5   EFFECT OF LATERAL PRESSURE ON BIAXIAL STRENGTH

8.5.1 Introduction

Dier and Dowling (1980) present a comprehensive study of the interaction of lateral pressure and biaxial in-plane compression. This study was also aimed at ship structures and the level of lateral pressure ranged from 0–0·1962 N/mm². Again, emphasis was on 3:1 plates, although square plates and plates of higher aspect ratio were also studied.

8.5.2 Interaction Results

Figures 8.13(a) and (b) show interaction curves of σ_x versus σ_y for plates with $b/t = 40$ and 60 ($\sigma_0 = 245$ N/mm²) and a 3:1 aspect ratio, and pressure heads of 0, 10 and 20 m of water. The figures also show the effect of edge clamping, which predictably for lateral loading has a very significant effect on strength. For high lateral pressure the effect of fabrication will be greatly enhanced by clamping induced by the pressure and in many instances it would seem justifiable to take a conservative value of the clamped condition curves.

The β-parameter can be used to non-dimensionalise panel slenderness

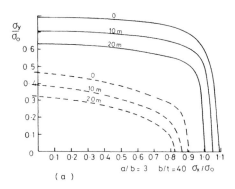

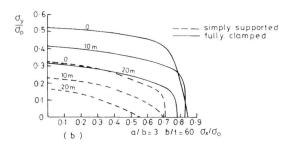

FIG. 8.13.   Effect of pressure on interaction curves for plates under biaxial compression.

with respect to yield as before, while the study shows that the non-dimensionalising factor for pressure is E/σ_0^2. For plates of different yield stresses therefore, the pressure loads quoted should be factored by $\sigma_0^2/245^2$.

The curves (Fig. 8.13) show that clamping has the highest effect for stocky panels as would be expected. This is because the benefit of clamping in slender panels is spread over a much smaller relative area of the plate and does not have as significant an effect on the buckling behaviour.

8.5.3 Effect of Imperfection

Figure 8.14 shows the effect of lateral pressure in reducing the imperfection sensitivity of plates subjected to biaxial loading. The 20-m pressure level effectively halves the imperfection sensitivity throughout the interaction range. This is not unexpected as plates are known to be less imperfection-sensitive under lateral loading than under in-plane compression.

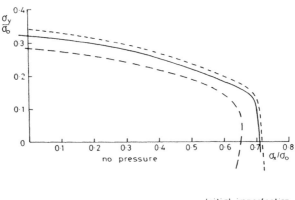

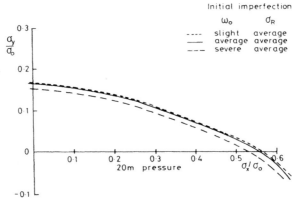

FIG. 8.14. Effect of pressure on the imperfection sensitivity of plates under biaxial compression ($a/b = 3$, $b/t = 60$).

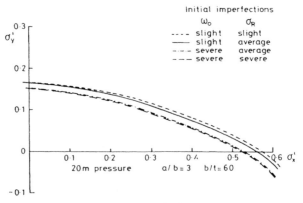

FIG. 8.15. Effect of imperfection and residual stress on plates loaded by biaxial compression and lateral pressure.

Imperfection levels of slight, average and severe quoted in this figure correspond to the equivalent imperfection levels considered in Section 8.4.

Figure 8.15 shows the extreme effect of varying both geometrical imperfection and residual stress level for the 20-m pressure level. It can be seen by comparing Figs. 8.14 and 8.15 that residual stresses have only a slight influence on plate panels subjected to high pressure and significant geometric imperfection.

8.6 PLATES UNDER SHEAR

8.6.1 Introduction

The discussion of this chapter so far has concentrated essentially on the interaction between yield and buckling for plates subjected to compressive stresses. While shear is a completely different loading type in many respects, the behaviour of a panel subjected to shear illustrates many of the features discussed in the previous sections.

It is important to note that shear is a loading mechanism that creates orthogonal levels of compression and tension in a panel. The resulting behaviour therefore comprises many of the phenomena relating to buckling in compression although most of the effects are moderated or softened to some degree by the presence of transverse tension.

The behaviour of a panel in shear undergoes three distinct phases, all of which contribute to its ultimate collapse strength. Figure 8.16 illustrates these phases.

The first phase occurs prior to significant out-of-plane deflection when compressive and tensile stresses are present to an approximately equal degree and the panel can be regarded as carrying its load by a pure shear-stress response. The point at which the transition from phase A to phase B occurs is dominated by the relationship between elastic critical stress and yield and never occurs for very stocky panels where the plate reaches a yield plateau under the action of the shear force.

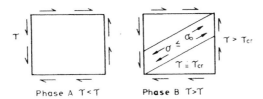

FIG. 8.16. Behaviour of a panel in shear.

The transition from phase A to B occurs when the plate is no longer able to sustain the diagonal compressive forces without buckling and a large growth in out-of-plane deflection results. Because of the stabilising effect of the tensile stresses the critical stress of a panel in shear is over twice that of a panel in compression, the buckling factor being 9·35 compared with a corresponding minimum compression factor of 4.

When buckling occurs, unlike in a compression panel, it occurs as a diagonal band. This is essentially because the line of the buckle opposes the compressive stresses causing it and also because the remaining diagonal tension pulls the lines of the plate relatively straight in the direction along which it acts.

As the loading increases beyond the critical load the compressive stresses cannot develop and the behaviour begins to be dominated by an increasing band of tension until this reaches a level which induces tensile yield. This band of tension is not necessarily along the panel diagonal but gradually orientates itself into a position maximising the shear resistance. The effect of this is studied in relation to tension field theories such as that described in Porter et al. (1975).

The final component of strength results from the frame action of the surrounding structure. In most practical girders, such as deep-plate girders with stocky flanges or wide-flange box girders, at least part of the flange is capable of acting with the edge boundaries of the web plate to provide a Vierendeel action which builds up as the vertical lozenging of the panel develops. This force component is included in the ultimate local calculations of Porter et al. (1975). It must be remembered, however, that the presence of direct stresses in the flanges will reduce the capability of these members to carry local moments needed in the Vierendeel action and this must be considered in design. As this chapter is concerned with plate panel behaviour the effect of this latter phenomenon on strength will not be discussed further.

8.6.2 Effect of Slenderness on Strength

Because of the presence of the diagonal tension band the effect of slenderness within the practical range of b/t values of between 60 and 240 is far less marked than for a compression panel.

For unrestrained edges, i.e. edges not carrying any transverse stress, the variation is significant and is shown in the upper part of Fig. 8.17. Tension field forces in this case cannot anchor transversely to the flange or adjacent panels and can only develop relatively small components within the panel itself. The results presented are from Harding et al. (1976) in which a strain

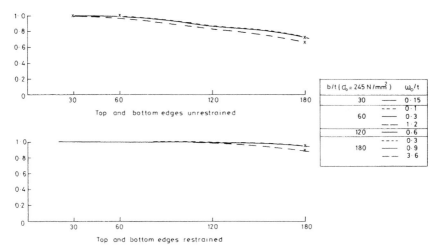

FIG. 8.17.   Effect of slenderness and imperfection level on shear strength.

control condition is applied along the line of the boundary to the panel representing the presence of a vertical stiffener, horizontal stiffener or flange. In the absence of other applied direct stresses the condition results in a zero tangential strain along all four boundaries thereby giving some degree of restraint to a tensile force parallel to the boundary itself.

The lower part of Fig. 8.17 shows that for restrained edges, capable of carrying net forces and remaining straight or with a fixed curvature appropriate to any other applied stresses, the dependence of strength on slenderness is much less. This is because the panel is capable of developing a large tension field force whatever its slenderness and this overcomes to a significant degree the effect of the strength limitations due to buckling.

8.6.3 Effect of Imperfection on Strength

Because of the presence of the tensile stresses the effect of imperfections, which has an influence only on the buckling behaviour, is relatively slight. Somewhat surprisingly, however, the effect of imperfections is relatively large for the case of the $b/t = 180$ panel with unrestrained boundaries shown in Fig. 8.17. This is mainly because of a limit on applied strain imposed when establishing these ultimate loads. A limit of $2\varepsilon_0$ (i.e. twice the shear yield strain) has been set when evaluating peak strengths in order, from a design point of view, to limit the overall panel-shear deformation and hence its potential destabilising influence on the compression flange or longitudinal stiffeners present in a box-girder design. With an imperfection of four times the thickness the peak strength occurs at significantly beyond the $2\varepsilon_0$ limit and the point shown on Fig. 8.17 is therefore less than the true maximum. It is true for the corresponding restrained results. It should be noted that the tolerance level thought appropriate for design corresponds to the central curve in the figure.

The effect of residual stress is also relatively little. This is discussed in more detail in the next section.

8.7   INTERACTION OF DIRECT STRESSES WITH SHEAR

8.7.1   Introduction

The main importance of the work of Harding *et al.* (1976) is the examination of the interaction between shear and in-plane compression, tension and bending stresses. The study was carried out to examine the likely load combinations found in the webs of box-girder bridges (intrinsically multiply-stiffened).

8.7.2   Stress–Strain Response

Figure 8.18 shows the direct and shear stress–strain responses of an unrestrained panel subjected to varying proportions of compressive and shear displacement. Six ratios of applied strain are considered and each ratio, with the exception of the shear strain-only and compressive strain-only cases, have two corresponding shear and direct stress–strain response curves. The responses are smooth without significant unloading, especially in the case of shear, because of the stockiness of the panel. The geometrical imperfection level considered is appropriate to a design tolerance although residual stresses are zero in this instance.

Curves of this type, with parametric variation, have been used to develop interaction diagrams between the various possible stress components. In doing so, because the work is related to a member primarily concerned with carrying shear, the peak of the shear diagram has been taken with the coincident direct stress. A strain limit of $2\varepsilon_0$ or $2\gamma_0$ has been applied to maintain overall stability. The interaction diagrams discussed in the next section should therefore not be used to assess the ultimate capacity of a panel primarily loaded in compression.

Figure 8.19 shows stress distributions and deflections for three strain ratios for a $b/t = 120$ panel under similar conditions to those of Fig. 8.18 for different levels of applied strain. The stress distributions are those appropriate to the left-hand panel boundary. Figure 8.19 is interesting

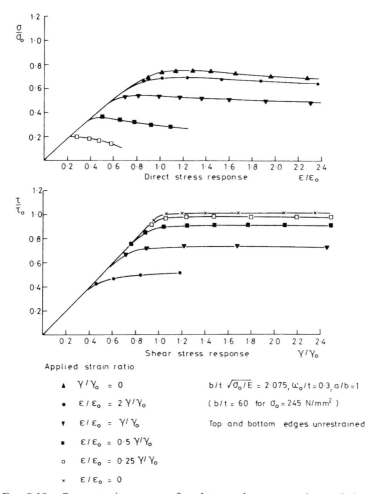

FIG. 8.18.   Stress–strain response for plates under compression and shear.

because it illustrates the change of mode from the pure compression to the shear cases. In the combined case the effect of the compression is still evident as deflections are significantly higher than the shear-only case and the out-of-plane curvatures are higher. The out-of-plane deflections produce very significant non-linearity of compressive stress for the $\gamma/\gamma_0 = 0$ case because of the relative slenderness of the panel in compression. The effect of the shear inducing a tension field stress is clear in the second diagram where the compressive stress is reduced to zero at the corner of the

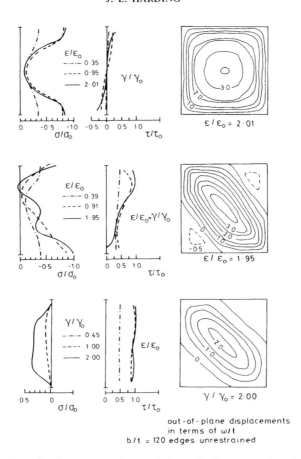

FIG. 8.19.   Growth of stresses and out-of-plane displacements for plates under shear and compression.

panel. Because of the concentration of the compression at the bottom of the panel, the shear distribution is also severely non-uniform with a shear 'boot' occurring at the left-hand top corner. For the shear-only case it can be seen that the shear-stress distribution is relatively uniform. This is essentially because of the effective stockiness of the panel under shear-only loading. The direct stress adopts a reasonably uniform tensile distribution in this instance caused by the out-of-plane deflections although the high corner tensile stresses are not present as the diagonal deflections are still only moderate due to the stockiness of the panel in shear.

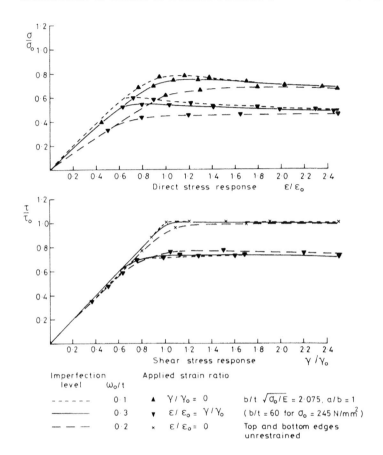

Fig. 8.20. Stress–strain response for plates under compression and shear—effect of geometric imperfection.

8.7.3 Variation with Initial Imperfection

Figure 8.20 illustrates the behaviour of a $b/t = 60$ panel under shear and compression with three levels of initial imperfection. The greatest variation is observed in the compressive response. It should be noted, however, that at the peak shear load, the imperfection sensitivity of the compression loads is very small as the curves tend to a limit once peak load has been passed. This should be considered, when examining the interaction curves presented in the next section.

As stated in the section on shear only, the apparent imperfection sensitivity to shear loading increases with slenderness due to the strain limit applied.

8.7.4 Interaction Results

The main purpose of this study was the production of interaction diagrams which could form the basis for the design of plate panels forming components of multi-stiffened web members.

The design application of the work is discussed in Section 8.8. This

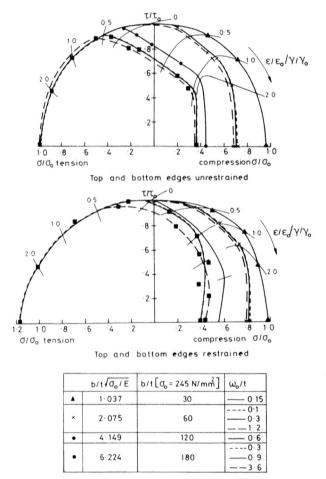

FIG. 8.21.   Interaction curves for plates under compression, tension and shear.

section deals with some of the phenomena affecting the interactive behaviour but, of necessity, not all cases can be presented.

The first figure (Fig. 8.21) shows the interaction of unrestrained and restrained panels under in-plane tension, compression and shear. It also indicates the imperfection sensitivity of the elements to these load types. No residual stresses are applied in this case. Several features of the diagrams are worth noting:

1. The unrestrained shear strength is much more sensitive to changes in panel slenderness than restrained strength which is insensitive.
2. The imperfection sensitivity of panels in both shear and compression is small considering the limitations imposed when formulating the diagrams.
3. Panels in tension are completely insensitive to imperfections and slenderness and, in the unrestrained cases follow the von Mises yield criterion until the ratio of shear to tension is high at which point buckling causes some reduction in strength.
4. Restrained panels in tension gain additional strength above nominal yield because of the interaction of transverse stresses.
5. Equivalent applied strain and stress ratios do not form coincident points on the interaction diagram. Because of the ability to shed stress, high slenderness panels, for example loaded by shear and significant tensile strain, might produce a stress-interaction position on the shear axis. This has implications on design which are discussed later.

Figure 8.22 shows the effect of residual stress level on the previous results. In this instance because of the presence of both flange/web boundaries, longitudinal stiffeners and vertical stiffeners in the type of structure being considered a biaxial stress state has been examined.

The level most appropriate to design is the 10% compressive stress in both directions and it can be seen from Fig. 8.22 that this level of residual stress does have a significant effect on strength for unrestrained panels, particularly in the slender range. The reason for this is again connected with the $2\varepsilon_0$ or $2\gamma_0$ strain limit imposed on panel deformation when deriving the interaction results. In the presence of biaxial residual stress the stiffness of the panel is greatly reduced from the onset of loading and the peak load occurs at very high strain levels. To some degree therefore the implications of the interaction diagram are false but account has to be taken of this lower strength at moderate strain levels in design.

In the case of the restrained results the reduction in strength is

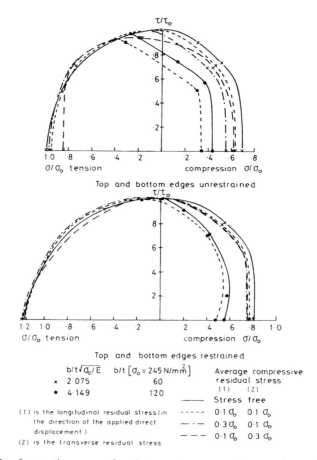

FIG. 8.22.   Interaction curves for plates under compression, tension and shear—
effect of residual stresses.

significantly less severe presumably because of the lower out-of-plane
deflections and therefore stiffer response, but it is still evident for slender
panels in regions of high compression. The ultimate shear load is almost
unaffected.

Figure 8.23 shows interaction curves between bending and shear. The
axis value M_0 is the elastic limit moment and M_u the fully plastic moment.
It can be seen that strengths significantly above M_0 are obtained in the
restrained case, even for panels with quite high slenderness, and this is
because of the capability to redistribute bending stresses to the centre of the

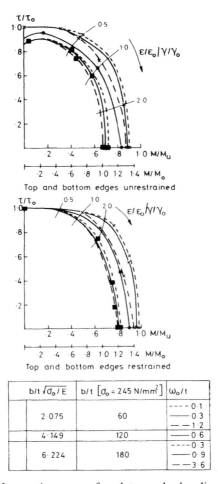

$b/t \sqrt{\sigma_0 / E}$	$b/t \left[\sigma_0 = 245 \text{ N/mm}^2\right]$	ω_0 / t
2·075	60	---- 0·1 —— 0·3 — — 1·2
4·149	120	—— 0·6
6·224	180	---- 0·3 —— 0·9 — — 3·6

FIG. 8.23. Interaction curves for plates under bending and shear.

panel once the edges have yielded. It is also apparent that quite high levels of shear are needed to significantly reduce the moment capacity of the panels, and to some degree the converse is also true. The linear type behaviour in the case of high slenderness panels in compression and shear is missing in this instance. It should be remembered that the critical buckling coefficient for bending is some six times that for compression and the effective slenderness is therefore much lower for the former.

Figure 8.23 also shows that bending/shear interaction is relatively insensitive to imperfection, again reflecting the relative stockiness of the panels and also the effect of taking coexisting stresses at the peak of the shear response.

8.8   USE OF THE RESULTS IN DESIGN

There are two main problems associated with the direct use of results such as presented in this chapter in design:

> The first is concerned with presentation. It is not generally feasible to present a large number of graphs related to varying parameters for a design application.
> The second relates to the conformity between the loading types and boundary conditions assumed in the analyses and those present in the actual structure.

In the studies described, load was applied by means of linear displacements that were kept proportional. The real loading situation is normally far more complex but is in general associated with proportional load control for the structure as a whole. Shedding can occur, however, in individual elements. In the case of box girders for example, shear forces and applied moments are related to the overall geometry of the structure although the shear and moment stresses carried by the web might be affected by redistribution.

The analysis method of applied displacement loading will undoubtedly allow a significant redistribution in individual calculations. This was mentioned in Section 8.7.4 in relation to the stress-interaction position corresponding to a given strain ratio. For example a slender panel under shear and applied compression will fail at an interaction position near the shear axis. This reflects the fact that a high slenderness panel can sustain only a certain low level of direct stress. For a panel with proportional stress loading to reach this equivalent position a degree of load shedding would therefore be implied. If the panel was unable to shed this stress because of the weakness of the surrounding structure it could only be designed for a lower collapse strength in shear approximating to the position of the required proportional stress loading on the interaction curve.

Using this approach it would seem reasonable to adopt a design philosophy allowing redistribution, where this is desirable, as long as the remainder of the structure is proportioned accordingly. As indicated above, one obvious case involves the shedding of direct stresses from the webs of a box girder to the flanges where they can be more efficiently resisted. It is of interest to note that in the new British steel bridge design rules 60 % of web direct stresses are allowed to be shed to the flanges. This applies to stresses from the entire web depth so that compatibility problems do not arise, although the sub-panels in tension will lose some shear-buckling resistance.

The boundary conditions of the analyses were, in general chosen to approximate those found in real structures. The study of Section 8.7 for example was performed with the aim of developing rules for the design of box-girder webs. Restrained boundaries were chosen to represent the stronger in-plane condition of internal panels and the unrestrained boundaries were chosen to represent external panels abutting flanges with weak bending resistance.

In the case of the compression study of Section 8.3, it was decided that the constrained-edge condition most closely modelled the situation within a multi-stiffened box-girder flange although, in the event, unrestrained and restrained conditions were chosen for the British code to corresponding internal and edge panels as for the web clauses.

The suitability of the two boundary classifications was checked to a limited extent using a discretely stiffened elasto-plastic large deflection computer program and this found the panel formulations to be appropriate as long as the stiffeners have sufficient axial and lateral stiffness to remain straight and apply a sufficient degree of strain control to the panel edges.

In applying the panel strength results to design it became apparent that a simplified interaction formula was needed to cover as general a loading case as possible, into which the computer results could be incorporated. To accomplish this a formula was developed, based essentially on elastic interaction but incorporating geometrical and material knock-down factors based on the numerical results.

The basic formula adopted for the combined compression, in-plane bending and shear was

$$\left[\frac{\sigma_c}{S_c\sigma_0}\right] + \left[\frac{\sigma_b}{S_b\sigma_0}\right]^2 + \left[\frac{\tau}{S_s\tau_0}\right]^2 \not> 1 \tag{8.1}$$

where S_c, S_b and S_s are knock-down factors from the analyses and the original version of these are given in Figs. 8.24(a), (b) and (c).

In establishing these, different aspect ratios were ignored for compression and bending because their effect on strength was only moderate, and a conservative approach was taken in formulating the curves. A third S_c-factor exists in the British design rules for long panels loaded across their width.

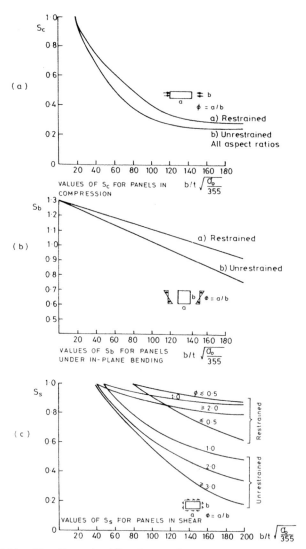

FIG. 8.24.   Non-linear buckling factors for use in interaction equation.

Tension can be allowed for by taking a -1 value for the S_c-term, allowing for the beneficial effect of tensile stress on buckling. It should be noted that a yield restriction is also needed as a separate limit and will restrict, for example, the benefit gained from the presence of a tensile stress in a stocky panel.

In eqn (8.1) σ_c, σ_b and τ are the components of compression, bending and shear applied to the sub-panel under consideration and σ_0 and τ_0 are the yield strengths in direct and shear stress.

The yield condition given in the British code corresponds essentially to

$$\left[\frac{\sigma_c + \sigma_b/1\cdot3}{\sigma_0}\right]^2 + \left[\frac{\tau}{\tau_0}\right]^2 \ngtr 1 \tag{8.2}$$

The factor $1\cdot3$ allows some peak overstress and recognises in part the tendency of peak-bending stresses to redistribute further down the panel.

A further complication was introduced into the British code to allow redistribution of direct stresses in a shear situation. This is in the form of a 'ρ'-factor which allows benefit to be gained from shedding stresses from restrained panels but not from unrestrained panels. The amount of shedding permitted is 60 % of the direct stress present, equally shed from all sub-panels to avoid compatibility problems.

The revised formula is

$$\left[\frac{\sigma_c}{(1-\rho)S_c\sigma_0}\right] + \left[\frac{\sigma_b}{(1-\rho)S_b\sigma_0}\right]^2 + \left[\frac{\tau}{S_s\tau_0}\right]^2 \ngtr 1 \tag{8.3}$$

where σ_c and σ_b now take the value of the reduced stress (e.g. $0\cdot4$ times their original value) and ρ has the value zero for restrained panels and the redistribution percentage (e.g. $0\cdot6$) for unrestrained panels.

Harding and Dowling (1979) give a fuller description of the background to the formula and its use in design. In applying this type of formula in design, load and material factors are needed as the calculation provides a direct estimate of collapse strength.

It is also worth noting that the formula can be enlarged, or the functions modified, to include the interaction of other stress types. For example biaxial compression can be included by modifying the first term to be

$$\left[\frac{\sigma_{c1}}{S_{c1}\sigma_0}\right] + \left[\frac{\sigma_{c2}}{S_{c2}\sigma_0}\right]$$

where σ_{c1} and σ_{c2} are the biaxial stress components and S_{c1} and S_{c2} are the S_c-functions in the two directions (taking the appropriate value of b/t).

A further allowance has been proposed for the inclusion of lateral loading where new S_c-functions are derived from the interaction results. Figure 8.25 shows a suggested set of values for clamped and simply-supported edges.

One minor problem with the equation is a degree of conservatism with respect to the interaction behaviour of stocky panels. A more complex formulation could be proposed but the existing formula remains attractive because of its simplicity.

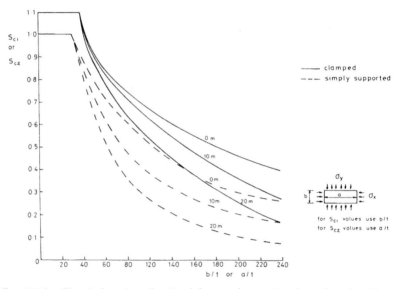

FIG. 8.25.   The S_c-functions for biaxial stress interaction including the effect of lateral pressure.

8.9   CONCLUDING REMARKS

This chapter has presented the results of computer studies on plate panels under various types of complex loading. It is intended to provide some understanding of the physical phenomena underlying the buckling problem as well as giving some guidance on the use of the work in the design situation.

REFERENCES

BLEICH, J. F. (1982) *Buckling Strength of Metal Structures*, McGraw-Hill, London.

CRISFIELD, M. A. (1975) Full-range analysis of steel plates and stiffened panels under uniaxial compression. *Proc. Instn Civ. Engrs*, December, 59.

DIER, A. F. and DOWLING, P. J. (1980) Strength of ships plating: plates under combined lateral loading and biaxial compression. CESLIC Report SP8, Imperial College, London.

DOWLING, P. J., FRIEZE, P. A. and HARDING, J. E. (1977) Imperfection sensitivity of steel plates under complex edge loading. *Conference on Stability of Steel Structures*, Liege.

DOWLING, P. J., HARDING, J. E. and SLATFORD, J. E. (1979) Strength of ships plating: plates in biaxial compression. CESLIC Report SP4, Imperial College, London.

FRIEZE, P. A., DOWLING, P. J. and HOBBS, R. E. (1977) Ultimate load behaviour of plates in compression. In *Steel Plated Structures*, Crosby, Lockwood and Staples, London.

HARDING, J. E. and DOWLING, P. J. (1979) The basis of the proposed new design rules for the strength of web plates and other panels subject to complex edge loading. In *Stability Problems in Engineering Structures and Components*, Applied Science Publishers Ltd, London.

HARDING, J. E., HOBBS, R. E. and NEAL, B. G. (1976) Parametric study on plates under combined direct and shear in-plane loading, CESLIC Report BG44, Imperial College, London.

LITTLE, G. H. and DWIGHT, J. B. (1972) Compressive tests on plates with transverse welds. University of Cambridge, Department of Engineering, Report CUED/C-Struct/TR31.

MOXHAM, K. E. (1971) Theoretical determination of the strength of welded steel plates in compression. University of Cambridge, Department of Engineering, Report CUED/C-Struct/TR2.

PORTER, D. M., ROCKEY, K. C. and EVANS, H. R. (1975) The collapse behaviour of plate girders loaded in shear. *Structural Engineer*, 53(8).

SMITH, C. S. (1981) Imperfection effects and design tolerances in ships and offshore structures. Paper No. 1434 presented at 1st Conference Inst. Engrs. and Shipbuilders in Scotland and RINA, Glasgow, *Proc. IESS*.

VALSGARD, S. (1978) Ultimate capacity of plates in biaxial in-plane compression. Det norske Veritas Report 78–678.

INDEX